Field Guide to

SEA STINGERS

AND OTHER VENOMOUS
AND POISONOUS MARINE
INVERTEBRATES
OF
WESTERN AUSTRALIA

Field Guide to

SEA STINGERS

AND OTHER VENOMOUS AND POISONOUS MARINE INVERTEBRATES OF WESTERN AUSTRALIA

LOISETTE M. MARSH and
SHIRLEY M. SLACK-SMITH
with medical contributions by DR DESMOND L. GURRY

Clay Bryce

Contents

Acknowledgements

We thank all those who have so generously shared information with us and those who have permitted publication of their slides or photographs; Dr Christopher Appleton who made his PhD thesis on Austrobilharzia (Murdoch University) available; Patrick Baker, Western Australian Museum, for photographs and a first-hand account of stinging by *Alatina* sp.; Harry Butler, conservation consultant, for a first-hand account of a stinging by *Echinothrix calamaris*; Dr Paul Cornelius, formerly of the Natural History Museum, London, for assistance with identification of jellyfish; Milton and Aileen East, Geraldton, Western Australia, for their graphic account of a stinging by *Conus geographus*; the late Dr Robert Endean, University of Queensland, for providing the case history of stinging by *Alatina* sp. at Exmouth; A/Prof Peter Fenner, James Cook University, Queensland, for assistance on numerous occasions; Dr Jane Fromont, Western Australian Museum, for providing information on toxic sponges; Dr Ray George, formerly of the Western Australian Museum, for his contribution on toxic crabs; Dr Lisa-ann Gershwin, Australian Marine Stinger Advisory Services, Townsville, and Dr Jacquie Rifkin for identification of jellyfish; Diana Jones, Western Australian Museum, for information on crustaceans and worms; George Kendrick, Western Australian Museum, and Dr Anne Brearley, University of Western Australia, for first-hand accounts of the results of stingings by *Conus anemone*; Dr Barry Russell, Museum and Art Gallery of the Northern Territory, for a first-hand account of stinging by *Alatina grandis*; Dr Geoff Taylor, Busselton, Western Australia, for information and photographs of *Alatina* sp. at Exmouth; U Weinreich, Townsville, Queensland, for information on the natural history of *Conus geographus*; St John Ambulance Australia for first aid procedures; Jill Ruse, Rachel O'Shea, and graphic artists, for the meticulous diagrams; Lee Stupart, editor; and Ann Ousey and Greg Jackson of the Publications Department at the Western Australian Museum, for their contribution to the production of this second edition.

Clay Bryce

◄ Coral reefs are beautiful but can hide many venomous and toxic animals.

Introduction

The coastline of Western Australia extends for over 12,000 km from the tropical waters in the north to the temperate waters in the south. Every year it attracts thousands of local, interstate and international swimmers, surfers, rowers, sailors, skindivers, fishers, naturalists, reef explorers and beach walkers.

In this book we aim to make everyone working or relaxing along the Western Australian coast aware that some of those animals in the sea (other than fishes and sea snakes) might cause injury.

We should all be mindful that the animals that live in the sea belong there, and that the venoms and poisons produced by some are normally used to deter predators or to subdue and capture their natural prey. Marine animals do not set out to injure or kill humans. However, by entering their territory we may appear as a threat and so we run the risk of being injured as the animals try to protect themselves.

We also aim to show that most of these dangers can be avoided by recognising those animals that might cause injury, by being aware of our surroundings at all times and by taking precautions such as wearing protective clothing.

In our descriptions of the animals we have used a few technical terms. A definition follows in brackets after the first time a technical word is used in the book, and that word is also defined in the Glossary.

Details on how to prevent injury are listed for each 'stinger' species or

Clay Bryce

◀ Coral reefs of the Houtman Abrolhos off Geraldton mark the southern limit of many of the more tropical venomous and toxic invertebrates in Western Australia.

group. We also have listed the symptoms and the treatment for injuries that could be caused by each 'stinger'. Such treatments are for mild and moderate injuries, and for first aid in severe cases. We stress the importance of seeking medical treatment for severe stings as quickly as possible.

It is interesting to note that in this time of intense biological and genetic research many of the toxic substances produced by marine invertebrates to capture food or deter predators are being analysed and assessed for therapeutic use in humans, and that these substances might eventually contribute to our wellbeing.

As a last word we think it's important to say that we have survived many years of scuba diving and most of our lives swimming and snorkelling in Western Australian and other seas with no more than very minor stings, and we believe our book will help you to do the same.

Loisette Marsh, Shirley Slack-Smith and Des Gurry

Clay Bryce

▲ Coral reefs whether covered in hard or soft corals provide many ledges and holes where venomous urchins and crown-of-thorns starfish hide during the day.

▶ The fire coral Millepora off Hibernia Reef, north-western Australia.

Clay Bryce

Sponges

Phylum Porifera

Sponges, despite their plant like growth and appearance, are animals, although some live in a symbiotic relationship with algae or cyanobacteria. They are individuals consisting of semi-independent cells that feed by straining minute planktonic organisms from the stream of water that enters through pores all over the body surface and leaves from one of the

FIGURE 1:

(a) A simple sponge cut to show the interconnecting canals, which carry water through it in the directions arrowed.

(b) Some examples of sponge spicules.

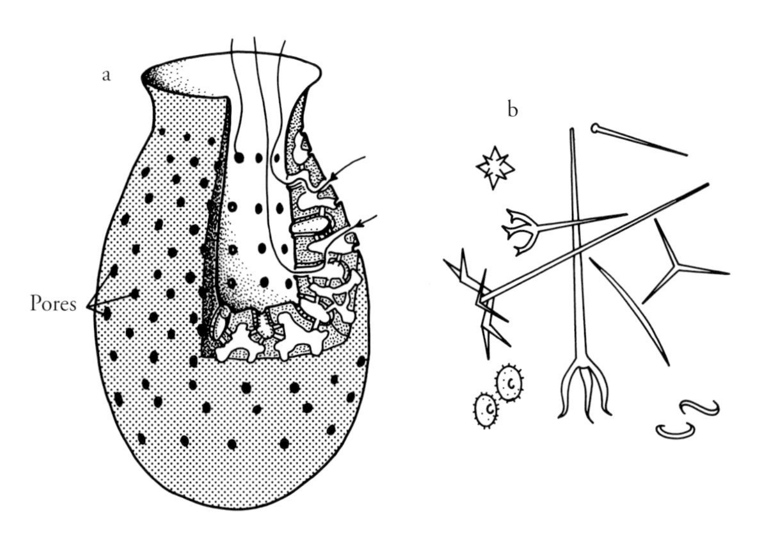

Clay Bryce

◀ Sponges and other filter feeders dominate the rocky subtidal sea bed off much of Western Australian coast.

larger holes at the top (Figure 1). Millions of pores give the sponges their scientific name Porifera, or 'pore bearers'.

Most sponges live in the sea, and range from those which have a definite shape to others which form irregular encrustations on rocks. Shallow water sponges are often brightly coloured, particularly the red, green, yellow, orange, pink or mauve sponges found under boulders or growing on the walls of undersea caves. Deeper water sponges are more often dull-coloured or white. They range in size from as small as a thumbnail to a metre or more in height. The sponge skeleton is a meshwork of horny spongin fibres strengthened in most by glassy spicules (or needles) of silica. However, the skeleton of spongin in soft bath sponges is a mesh of soft horny fibres without spicules.

Sponges, alive or dead, can cause irritation in different ways, each requiring different treatment. The needle-like glassy spicules can pierce the skin when the sponge is handled, or can enter the eyes of fishers as they haul up deep water trawl nets. The spicules can also be inhaled or swallowed after the sponge fragments have dried out, particularly those from freshwater sponges in drying ponds.

Sponges contain a variety of toxic compounds to deter potential predators and the slimy secretions from some cause severely painful contact dermatitis. The chemicals may also be damaging to the eyes. In addition, creatures living on and in sponges (e.g. anemones and bacteria) may cause infections or dermatitis.

Toxic sponges

Class Demospongiae

Family Desmacellidae

In 1991, a species of toxic sponge, *Biemna saucia* (below), was described for the first time from the north of Western Australia and the Northern Territory. It has been found on subtidal sandy rock or dead coral reefs at 6–20 m depth and has also been dredged from 74 m. At present it is known from north of Port Hedland, Western Australia, to Arnhem Land, Northern Territory (Hooper, Capon & Hodder 1991).

John Hooper

▲ A northern Australian sponge, *Biemna saucia*, which is easily fragmented, releasing an irritating secretion.

In colour *Biemna saucia* ranges from nearly white to yellow-brown. The sponge forms a soft mass, and attaches itself to rock which may become buried under the sand. It has erect fibrous digits 9–180 mm long and up to 80 mm thick, each with branches which may fuse between digits giving the sponge a shaggy, open net-like consistency. It's skeleton is composed of bundles of spicules loosely held together by very little spongin, so that the sponge may be easily broken and release spicules and toxic secretions into the water. The toxic metabolite (p-hydroxybenzaldehyde) is thought to be the cause of severe dermatitis, which may result from handling the sponge or by coming into contact with the toxic secretion.

Ron Southcott

▲ This southern Australian sponge, *Neofibularia mordens*, causes persistent skin rashes when handled.

In the same family of sponges, a highly toxic sponge *Neofibularia mordens* (page 10) has been found in South Australia and may also occur on the south coast of Western Australia. It is a massive sponge, as much as 40 cm high and wide. When alive, it is royal blue with a smooth slimy surface but, out of the water, its colour turns to a dark bluish chocolate brown. Its cut surface is light brown with a gritty texture. Contact with the slime causes intense, long-lasting itching. Reaction due to *Neofibularia mordens* is the most severe of any toxic sponge so far recorded from Australia (Southcott & Coulter 1971).

Toxic sponges of several other families known to occur in Western Australia include an Indo-Pacific species *Iotrochota baculigera* (Family Iotrochotidae) found from the Houtman Abrolhos to Darwin, and the Mediterranean and Indo-Pacific species *Tedania anhelans*, also known as *T. digitata* (Family Tedanidae). *Tedania anhelans* has been recorded from all areas of the Australian coastline and in Western Australia has been found in waters off Albany, Geographe Bay and Cockburn Sound to Shark Bay and the Dampier Archipelago.

Toxic sponges known from other parts of Australia but not yet having been found in Western Australian waters are *Neofibularia irata* (Family Desmacellidae) and *Spirastrella inconstans* (Family Spirastrellidae), both known from the Great Barrier Reef, Queensland, and *Lissodendoryx* sp. (Family Coelosphaeridae) from South Australia.

Since many reports of injuries from toxic sponges are anecdotal, it is rarely possible to relate them to particular species.

INJURY PREVENTION

- Wear a wetsuit and gloves when diving.
- Never handle a sponge, dead or alive, without gloves.
- Avoid touching or sitting where sponges have been lying.
- Use safety glasses when emptying trawl nets.
- If there is any possibility of sponge contact with bare hands, do not then touch delicate skin on face, eyes, genitals etc.

SYMPTOMS — FROM CONTACT WITH *BIEMNA SAUCIA*

- The secretion, which may contain spicules, can even penetrate lycra, thin wetsuits and gloves, and may cause a sting lasting for up to two hours.

- Itching, burning and sometimes redness after some time, followed by swelling, pain and a feeling that the skin has been scraped and even that there is glass in it.
- Mild toxic reactions may follow disturbance to the sponge without actual contact.

SYMPTOMS — FROM CONTACT WITH *NEOFIBULARIA MORDENS*

- Intense itching pain which is resistant to all attempts at relief. It often starts hours after touching the sponge and lasts for days. There may be little to see on the surface of the skin which is hurting so much.
- In other cases, an obvious itching and swelling redness of the skin (urticaria) may develop, and symptoms might last for weeks, followed by peeling of the skin

SYMPTOMS — FROM CONTACT WITH NON-TOXIC SPONGES

- Could leave tiny glassy spicules in the skin, which may become very painful.

SEVERE SYMPTOMS OF NON-TOXIC SPONGES

- If material is inhaled and lung irritation appears — causing breathing difficulty, wheezing and possibly coughing of blood — seek medical treatment urgently.
- If the material is swallowed there could be internal bleeding, for which there is little direct treatment.
- Severe gastric or intestinal bleeding could produce black stools (melena) which requires immediate medical attention.

TREATMENT

The symptoms will abate with the removal of foreign matter. The long duration of symptoms is indicative of the ineffectiveness of most treatments but the measures below may give relief.

- Remove visible glass spicules.
- Apply adhesive tape and pull off to help remove invisible glass spicules.
- Gently wash the contact area with soap and fresh water.
- Do not rub.

- Itching pain is not easily relieved or remedied but applying calamine may give relief in some cases.
- Cold therapy may ease pain. Wrap solid ice or cold packs used in eskys with layer of towel (to prevent severe cooling) and apply. Crushed ice in a plastic bag and cool cans can be safely applied to the stung area. Assess pain at 15 minute intervals and reapply cold pack if necessary.
- Avoid aspirin for pain relief, if spicules have been inhaled or swallowed, because it is an anticoagulant (allows bleeding to continue).
- Medical treatment might include:
- Application of burn cream such as 1% butesin picrate for dermatitis caused by *Biemna saucia* contact.
- Application of ultra-potent local steroid cream to the contact area, but not on subsequently infected lesions.

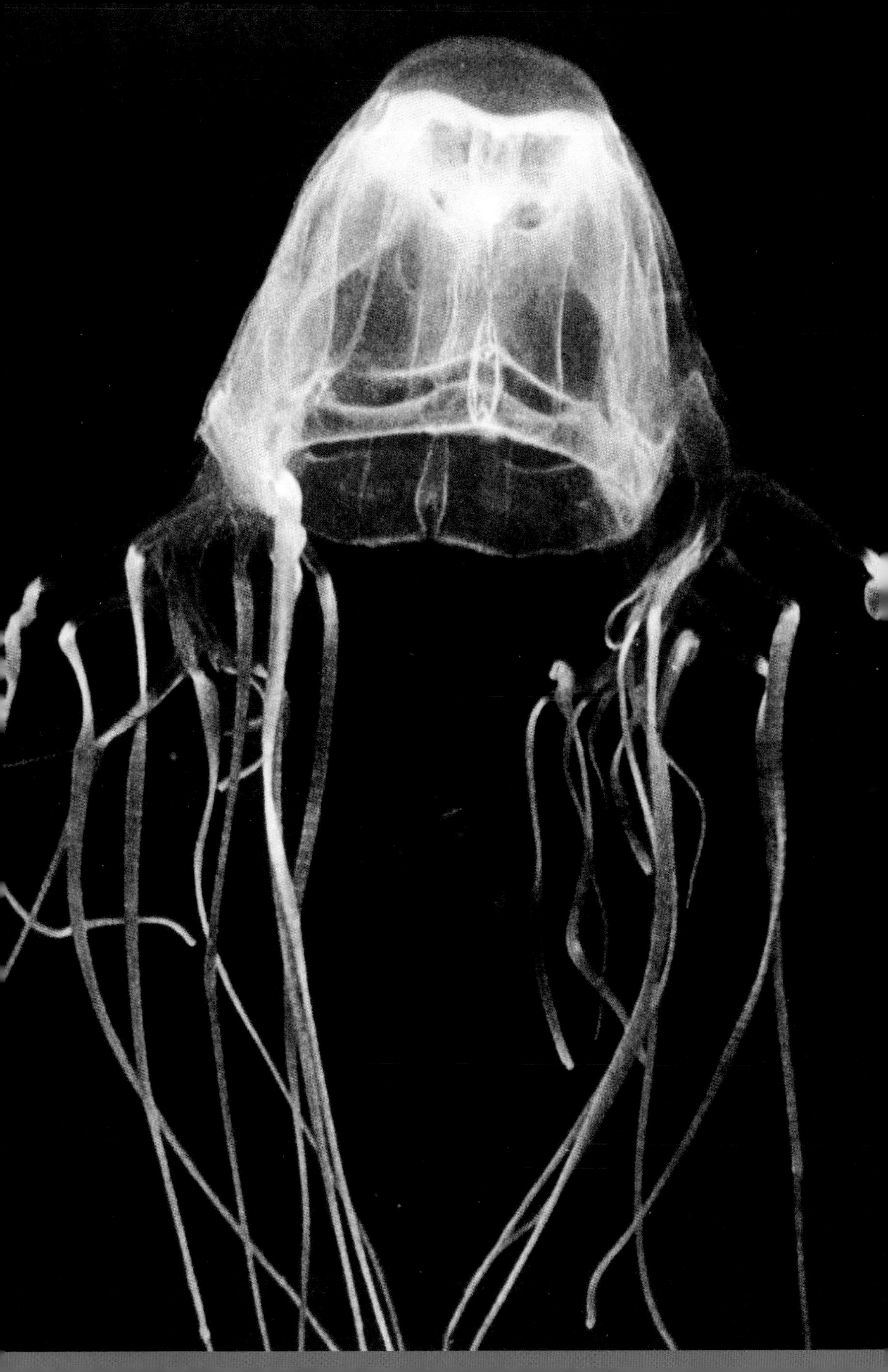

Jellyfish, Sea anemones, Corals and their relatives

Phylum Coelenterata (also known as Cnidaria)

Jellyfish, sea anemones, corals and their relatives belong to the biological group, or phylum, known as *Coelenterata* (and as *Cnidaria*). Many coelenterates look like plants but are really simple animals. They have a basic body plan with a body wall surrounding a central cavity with a single aperture, the mouth, that is surrounded by tentacles. The body wall has a gelatinous middle layer that makes up the bulk of the body in jellyfish.

FIGURE 2:

A polyp and a medusa – variations on the coelenterate body form.

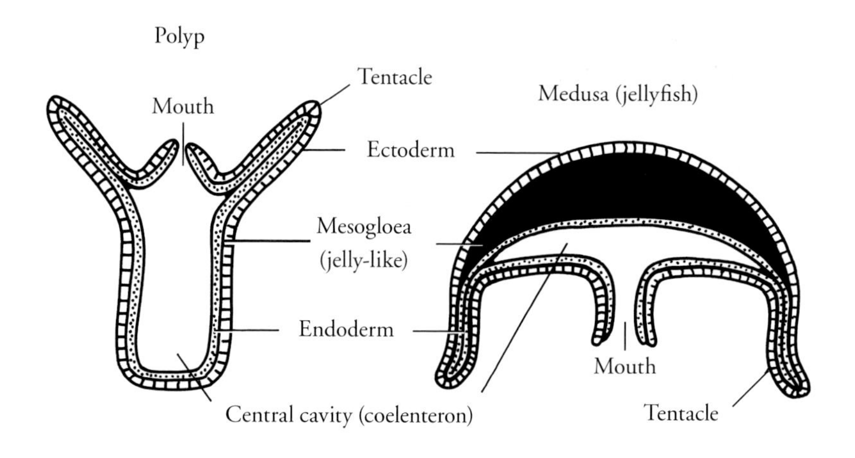

John Williamson

◀ The deadly box jellyfish *Chironex fleckeri*. See page 55.

The large group of animal species that make up the phylum Coelenterata is divided into four smaller groups, or classes: Hydrozoa (sea ferns, fire-weeds, hydromedusae, fire corals and bluebottles), Cubozoa (box jellyfish), Scyphozoa (round jellyfish) and Anthozoa (sea anemones, corals and soft corals).

Some of these animals live singly but many are joined to other individuals to form complex inter-connecting structures. A coral colony or a bluebottle are typical examples.

The body form of a coelenterate species may be a polyp (anemone type), a medusa (jellyfish type), or both (Figure 2). Hydrozoans usually have a life cycle involving both types: small floating or planktonic medusae produce eggs or sperm, which after fertilisation grow into sedentary polyps. These polyps then reproduce asexually by simply dividing but not separating, to form colonies that are often plant-like in appearance. Some of these polyps specialize to bud off minute medusae. In cubozoans and scyphozoans, the polyp stage is very small and the medusa is the conspicuous stage of the life cycle. Anthozoans form more complex polyps that reproduce both sexually and asexually without going through a medusa phase.

THE STING

Coelenterates catch their prey and defend themselves by using stinging capsules called nematocysts, which are located on the tentacles. The nematocysts are so minute that each one lies within a cell. However, groups of them may be visible as beads or rings on the tentacles of jellyfish or bluebottles. There may be up to 1,500 nematocysts to a square millimetre of the long tentacles of the deadly box jellyfish, *Chironex*.

Each minute nematocyst or stinging capsule contains a coiled thread that shoots out by reflex action when its trigger is either touched or stimulated by certain chemicals. The thread pops out suddenly like the long finger of a rubber glove being turned inside out. Some nematocysts shoot out long closed tubes that stick to the skin of the prey or entangle and anchor it. Others contain open-ended tubes that are able to penetrate the skin and inject a polypeptide venom. By absorbing water from the outside the nematocyst acts like a self-loading syringe and injects a volume of liquid more than 10 times the original volume of the cyst. As the prey struggles, more nematocysts are triggered to discharge and the venom paralyses the

FIGURE 3:

The outer cell layer of a jellyfish tentacle with an
(a) undischarged nematocyst
(b) one partly discharged
(c) and a fully discharged nematocyst with its hollow barbed thread through which venom is forced into the prey.

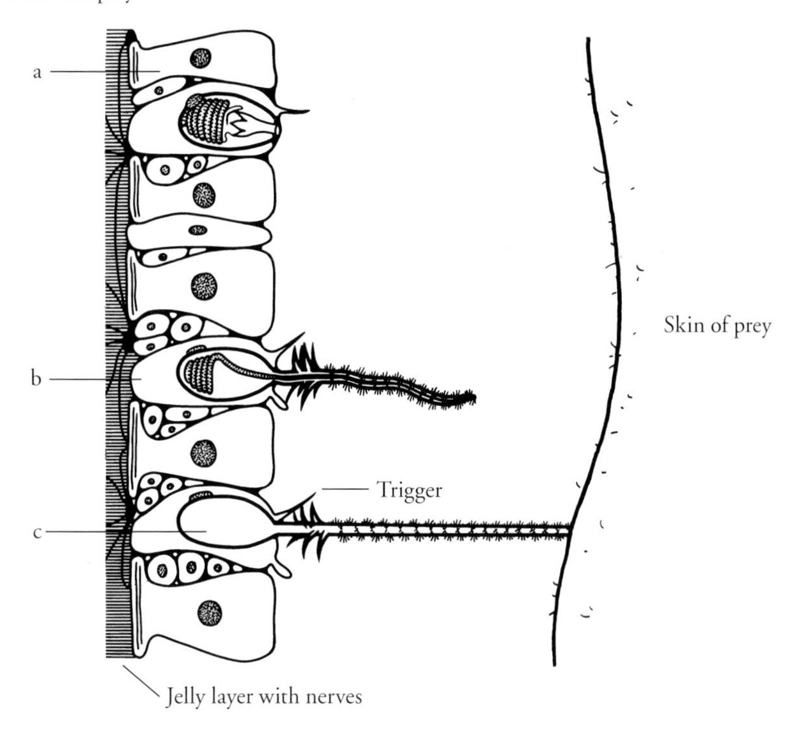

prey (Figure 3) so that the soft coelenterate is not damaged itself.

Each species of coelenterate has several characteristic types of nematocyst so that stinging capsules recovered from the skin of a victim can help to identify the stinger. Their shape, the length and arrangement of spines on their threads, the type, number and pattern of stings are also useful guides to the identity of the stinger. The type of venom also varies from one species to another and methods of deactivating of nematocysts can vary from species to species (Cleland & Southcott 1965).

In areas where stings are severe and well known, e.g. Queensland, lifesavers use adhesive tape to peel nematocysts off the stung area to use as samples for the microscope — a valuable step for diagnosing in cases of mass stinging. A full discussion on the structure and function of nematocysts, with identification keys, can be found in Williamson *et al.* (1996, pp 155–73).

Contact with and injection by the nematocysts of most coelenterates, such as many of the anemones living on rocky reefs, merely brings an awareness that they feel sticky to the touch. However, a few species of box jellyfish possess nematocysts that inject venom directly into the bloodstream so that it spreads rapidly and with great effect from the contact area to distant parts of and even to the whole body of a victim.

The severity of stings can vary depending on the nature and strength of the toxin, the hunger of the animal, the concentration of nematocysts on the tentacles, the surface area of contact, the duration of contact and the

Clay Bryce

▲ Nematocyst beads on a tentacle of a bluebottle *Physalia utriculus*.

resistance of the skin of the victim. For example, a small mildly venomous jellyfish brushing briefly over the hand of an adult male swimmer may be barely noticeable in its effect while the same jellyfish may cause very painful stings to the tender skin of a child or to soft skin such as that in the armpit, torso or groin, particularly if the jellyfish is trapped under clothing. Multiple stings from a swarm of jellyfish may cause severe injury and general illness. Children are more vulnerable than adults, and women more than men, due to more delicate skin and less body hair.

PREVENTION OF STINGS

Since most jellyfish occur at very sporadic and unpredictable intervals it is important to minimise the risk by wearing protective clothing such as a lycra stingersuit or a wetsuit whenever snorkelling or diving, particularly in tropical waters. This will also give protection when brushing against fire coral and stinging hydroids on coral reefs, and protect against coral scratches, which often become infected. When reef walking, it is advisable to wear strong canvas boots or thick-soled dive bootees with ankle protection.

Nematocysts adhering to and drying onto nets or ropes can sting many months later when rehydrated by wet hands, so wear gloves when handling nets or ropes that may have been contaminated.

Jellyfish do not sting deliberately; they use their stings for capturing food or for defence and only accidentally affect humans, usually through our lack of knowledge or carelessness.

TREATMENT OF COELENTERATE STINGS

Deactivating nematocysts to prevent further stinging is the first action to take in treating a coelenterate sting. It is important to assume that any piece of a coelenterate stuck to the victim's body will have discharged only part of its armoury and that undischarged nematocysts are present in the contact area either in pieces of tentacle or in isolated cells. Therefore, before removing any visible piece of the coelenterate from the victim's body, take steps to deactivate all undischarged nematocysts.

Treatment is also aimed at neutralizing the venom already injected and relieving pain that results from injected venom. Unfortunately nematocyst venoms are not uniform in their nature or response to chemicals so few generalisations can be made. However, all box jellyfish nematocysts are deactivated by household vinegar (4–5% ascetic acid). Vinegar is also effective against the bluebottle (*Physalia utriculus*), but causes further discharge of nematocysts in the multi-tentacled men-o'-war (*Physalia* spp.) and all other coelenterate stings. It is important to note that vinegar does not reverse the effects, including pain, caused by venom already injected.

To date, a specific antivenom has been developed against only one jellyfish, *Chironex fleckeri*, the most dangerous box jellyfish. First aid treatment of severe stings from *Chironex fleckeri* requires immediate dousing with vinegar, cardiopulmonary resuscitation (CPR) if it becomes necessary and medical treatment as quickly as possible.

Traditionally, alcohol such as in methylated spirits or concentrated drinks (whisky, vodka, etc) was generally used for marine stings but it is no longer recommended. In one study (Fenner & Fitzpatrick 1986) methylated spirits did not cause discharge of nematocysts in a round jellyfish stinger but in another study (Rifkin 1996), it caused massive discharge. Further, alcohol causes immediate discharge of box jellyfish nematocysts and so should **never** be used for stings from box jellyfish such as *Chironex fleckeri*.

Heat and cold both play a part in treating the pain from a coelenterate sting. Most marine venoms are destroyed by heat, at temperatures over 43°C. Cold relieves inflammation and pain but does not destroy the venom or the nematocysts, which tend to be preserved by cold. Obviously hot and cold cannot be used together; the choice, or order of application, would depend on what is available and safe.

Hot water treatment is effective for stings from both large and small four-tentacled box jellyfish. It has been tested on *Carybdea xaymacana* and *Alatina* sp. and is probably effective for other species. For treating with heat, immerse the stung area in water as hot as can be borne. Avoid scalding by using an unstung area or getting a helper to test the temperature of water. Fresh water, unless hot, should not be used as it causes immediate discharge of nematocysts. Treating with heat destroys the venom, bringing fast relief from most jellyfish stings (Taylor 2000). The higher the temperature of the water, the more rapidly the venom is destroyed (Carrette 2002).

Pain relief by applying ice, icepacks, and cold bottles and cans — wrapped in a cloth to avoid skin damage — is temporary, and pain may return as the coolant warms up. If this occurs, the cold treatment can be repeated for as long as required (Exton 1989). Cold therapy can be used for all jellyfish stings, and is widely used as a simple, safe and effective measure for pain relief.

Local anaesthetics such as a lignocaine spray (2% dilute) and ointment, both available from pharmacies over the counter, can be used, as can general analgesics such as paracetamol, aspirin and codeine.

Compression bandaging, which is so valuable in the treatment of snakebite, is included in some first aid protocols for major jellyfish stings. However, if applied before nematocysts have been deactivated, compression bandaging could make more nematocysts fire off into the flesh and allow more venom to reach the bloodstream more quickly. With life-threatening stings such as those from *Chironex fleckeri*, compression bandaging could hamper first aid priorities which are cardiopulmonary resuscitation, and access to antivenom and pain relief.

Results from independent tests (Hartwick, Callanan & Williamson 1980a, 1980b) indicate that applying aluminium sulphate (20% suspension), marketed as a spray or cream for pain relief, neither inhibited nor caused nematocyst discharge in the dangerous box jellyfish, *Chironex fleckeri*. It did not cause nematocyst discharge in the Pacific man-o'-war (*Physalia* sp.) or hairy stinger (*Cyanea* sp., also know as 'snotty').

Sodium bicarbonate (baking soda) is a mild alkali and Burnett's detailed laboratory tests showed that, as a powder sprinkled on the wet skin to make a slurry or thin paste it was the best deactivator of the stinging cells of the sea nettle, *Chrysaora* spp., a round jellyfish often found off metropolitan beaches in Perth, Western Australia.

Folk and traditional remedies include using the juice of a green pawpaw (in the Philippines) and the bruised leaves of the morning glory *Ipomoea brasiliensis* (once called goat's foot or pes-caprae), a creeper found on almost every tropical Indo-Pacific beach (Pongrayoon, Bohlin & Wasuwat 1991). Northern Australian Aborigines use the latex (white sap) from the cut young stems of beach morning glory (Cribb 1985) and elsewhere in the tropics the heated leaves are used externally as a painkiller. In north Queensland the sap of the crinum lily, *Crinum pedunculatum*, is used as a local anaesthetic as, in the eastern states of Australia is, the pigface, *Carpobrotus glaucescens* (Cleland & Southcott 1965; Cribb 1985).

Other traditional methods to stop further stinging and pain include pouring oil or sprinkling hot dry sand onto the skin as a barrier. Do not use wet sand and never rub the skin.

Extract of papaw (papain, 12 mg/g) is marketed as a cream to relieve pain and itching from bites and stings including jellyfish stings. It does inhibit nematocyst discharge in the saucer jellyfish, *Chrysora* sp., but does not relieve the pain and may damage the skin (Burnett *et al.* 1983b).

Aluminium chlorhydrate in deodorants has been said to relieve pain.

Part of the damage from a sting may be due to an allergy or hypersensitivity on the part of the victim, shown locally by weals and itching and, more remotely, by symptoms such as wheezing, coughing and sneezing.

As the foregoing illustrates, few generalisations can be made for treating coelenterate stings. A summary of current recommendations for treating mild to moderate coelenterate stings follows, and specific treatment for stings from a particular species or group of species is described where applicable. The awareness and recognition of likely stingers both provide valuable information for selecting the optimum treatment. Treatment of all severe coelenterate stings requires medical treatment as quickly as possible.

The beach morning glory *Ipomoea brasiliensis* (formerly known as *I. pes-caprae*) is a creeper found on almost every tropical Indo-Pacific beach. An extract of the leaves has been tested in Thailand and was found to be 80–100% effective against the proteolytic and haemolytic effects of *Cyanea* venom. It is now marketed as a cream with 1% IPA (extract of *Ipomoea*) but has not been tested on *Chironex* stings.

(Pongprayoon *et al.* 1991)

TREATING MILD TO MODERATE COELENTERATE STINGS

The first action is to deactivate the nematocysts, (present either as invisible loose cells or in obvious pieces of tentacle), and so prevent further stinging.

VINEGAR FOR DEACTIVATING STINGS FROM ALL BOX JELLYFISH AND THE SINGLE-TENTACLED BLUEBOTTLE (SEE PAGE 31, 39–67)

- Douse with vinegar in large quantities for over 30 seconds.
- Do not use alcohol such as in methylated spirits or concentrated drinks (whisky, vodka etc) as this activates nematocysts in box jellyfish causing further stinging.

HEAT FOR DEACTIVATING STINGS FROM LARGE AND SMALL FOUR-TENTACLED BOX JELLYFISH

- Soak stung area in water as hot as can be borne without further injury or scalding; use an unstung area to test temperature. Repeat as water cools. Hot water soaking works fast and if successful, there will be no need for any further treatment.
- Never use fresh water unless it is hot.
- Never rub or wash the skin vigorously, which could spread the nematocysts and, by applying pressure, cause nematocysts to discharge.

BICARBONATE OF SODA FOR DEACTIVATING ROUND JELLYFISH STINGS

- Apply a thin paste of bicarbonate of soda for sea nettle (*Chrysaora* spp. page 74) stings (may also help for mauve stinger, *Sanderia* sp. and hairy stinger or snotty).
- Do not use vinegar, which could enhance stinging activity of nematocysts in round jellyfish.

  Remove obvious sticky pieces of animal using tweezers, a stick, a comb, etc. Be aware that loose nematocysts invisible to the eye will be sticking to the skin.

PAIN RELIEF

- For most jellyfish stings a cold pack is the most effective measure for pain relief.
- Wrap solid ice or frozen gel bag with layer of towel (or cooling could be too severe) and apply. Crushed ice in a plastic bag or cool cans can

be safely applied to the stung area. Assess pain at 15 minute intervals and reapply cold pack if necessary.

- Aluminium sulphate as an ointment or spray may ease symptoms for stingings over a small area, and so may aluminium chlorhydrate, found in deodorants.
- Applying the juice of a green pawpaw or the crushed leaves of the tropical beach morning glory may help.
- Keep stung area rested and if possible and relevant, elevated.
- For persisting pain, use a local anaesthetic such as lignocaine as an aerosol, gel or ointment, or a general analgesic such as paracetamol, aspirin or codeine.
- If there is no relief from pain or other symptoms develop, seek medical attention.

FIRST AID FOR SEVERE STINGS FROM BOX JELLYFISH

e.g. ***Chironex fleckeri*** and Irukandji stingers

- Apply vinegar liberally for at least 30 seconds to deactivate nematocysts.
- Apply cardiopulmonary resuscitation if needed (see pages 229–231).
- Get the patient to hospital urgently.

Hydroids

FIRE-WEEDS OR STINGING HYDROIDS

Phylum Coelenterata
Class Hydrozoa
Order Hydroida

The fire-weeds, or stinging hydroids, are delicate fern-like colonies of animals, from a few millimetres to 20 centimetres in length, living on rocks and jetty piles. Each tiny branch bears many minute polyps, each with tentacles armed with stinging capsules or nematocysts used for food gathering.

Many hydroids have nematocysts capable of penetrating and stinging human skin. However, two species are particularly venomous and are known as fire weeds. The feather fire-weed, *Aglaophenia cupressina*, resembles brown seaweed and grows in clumps with flat fronds of about 15 cm in length. *Macrorhynchia philippina* (formerly *Lytocarpus*

Clay Bryce

▲ Fire-weed or Stinging hydroid, *Aglaophenia cupressina (Left)* showing capsules in which reproductive Medusae develop. *Macrorhynchia philippina (Right).*

philippinus), resembles a small white bush of delicate fronds with many branches arising from a single horny base. It is conspicuous on coral reefs, particularly along reef edges and channels.

Both species occur in north-western Australia, usually on coral reefs although occasionally found in shallow water as far south as Fremantle. Fireweeds can grow on ropes of buoys, jetties, etc.

INJURY PREVENTION

- All hydroids should be treated with caution as potential stingers.
- Wear gloves and protective clothing while diving or snorkelling, or handling mooring ropes.

SYMPTOMS

- From a mild stinging sensation to extreme pain, usually increasing over the first ten minutes.
- Pinpoint lesions may follow, developing into a generalised blotchy red rash, with swelling and sometimes blister formation. Rash may last up to ten days with local discomfort and itching.
- Generalised reactions are uncommon but may include abdominal pain, cramps, nausea and diarrhoea.

TREATMENT

- Wash with vinegar to deactivate nematocysts of *Macrorhynchia* and prevent further stinging; more research is required to determine if vinegar is effective for stings from other species.
- Do not use alcohol (such as methylated spirits or concentrated drinks such as whisky, vodka etc.) or fresh water.

Refer to treating mild to moderate coelenterate stings, pages 23–24.

HYDROMEDUSAE

Phylum Coelenterata

Class Hydrozoa

The medusae of hydroid polyps are known as hydromedusae, and are the sexually reproductive stages of hydroid polyps. They are mostly small round jellyfish and differ from Scyphozoa and Cubozoa in usually having four, often undivided, radial canals through the 'bell' connecting to a marginal circular canal. In some species, additional short canals extend from the marginal canal into the jelly layer.

Olindias phosphorica (Delle Chiaje, 1841)

Phylum Coelenterata

Class Hydrozoa

Order Limnomedusae

Family Olindiadidae

The small medusa of *Olindias phosphorica* (below) has a colourless, firm, hemispherical bell 1–3.5 cm in diameter. It has a central mouth and four radial canals, each with a frilly pink gonad attached along part of

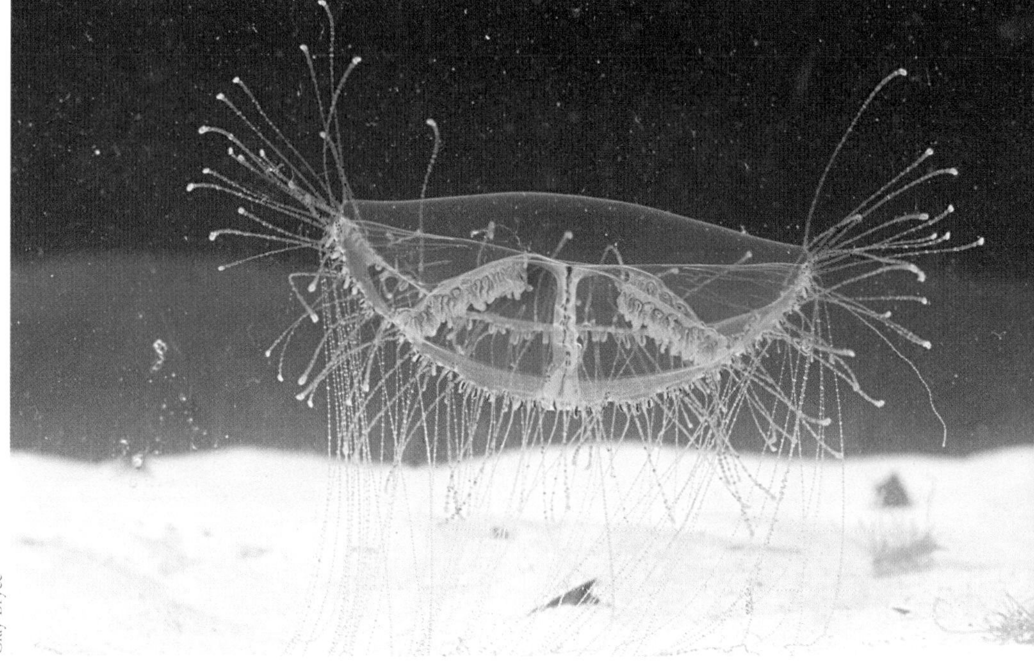

Clay Bryce

▲ *Olindias phosphorica.*

its length, and several short unbranched canals running inwards from the margin. It has a fringe of short marginal tentacles and many longer tentacles which, when contracted, are marked with brown beads or rings of nematocysts (stinging capsules).

The medusa of this species is the sexual stage of a small polyp that lives attached to hard underwater surfaces such as jetty piles and rocks. It is found in sheltered sandy bays, at 0–15 m depth.

Olindias phosphorica occurs in the tropical Indo-West Pacific region and is known from the coasts of Western Australia, South Australia, New South Wales and Queensland. The medusae are most commonly seen in late summer.

A second species, *Olindias singularis*, has a similar distribution, and gives a painful sting.

INJURY PREVENTION

- Wear a lycra stingersuit or wetsuit when snorkelling or diving.
- Keep a look out to avoid jellyfish.

SYMPTOMS

- A mild or sharp sting, which lasts about 30 minutes.
- Faint pink weals form after about 15 minutes and last for about two hours.

TREATMENT

- Wrap solid ice or frozen gel with layer of towel (or cooling could be too severe) and apply. Crushed ice in a plastic bag and cool cans can be safely applied to the stung area. Assess pain at 15 minute intervals and reapply cold pack if necessary.

Refer to treating mild to moderate coelenterate stings, pages 23–24.

Little red stinger

Leuckartiara gardineri Browne, 1916

Phylum Coelenterata
Class Hydrozoa
Order Anthomedusae
Family Pandeidae

The little red stinger, *Leuckartiara gardineri*, is a small hydromedusa only 10 mm high by 5 mm across its conical bell. The bell, clear at the apex, is filled by the orange reproductive organs, or gonads, and the large, red, folded mouth. Four pink tentacles, each about 25 mm long, together with numerous minute fine tentacles hang from the edge of the bell.

Since it is known from the Indian Ocean and north Queensland waters, *L. gardineri* could be expected to occur along the Western Australian coast, particularly in the north.

The relatively harmless *L. gardineri* could be confused with the more dangerous Irukandji stingers (refer pages 44), which are similar in size but have the four tentacles borne on bases called pedalia and have no fine marginal tentacles.

INJURY PREVENTION

- Wear protective clothing when snorkelling or diving.

SYMPTOMS

- Sharp sting that rapidly subsides.

TREATMENT

- Wrap solid ice or frozen gel with layer of towel (or cooling could be too severe) and apply. Crushed ice in a plastic bag and cool cans can be safely applied to the stung area. Assess pain at 15 minute intervals and reapply cold pack if necessary.
- Do not use vinegar or fresh water.

Refer to treating mild to moderate coelenterate stings, pages 23–24.

BLUEBOTTLES AND MEN-O'-WAR

Physalia spp.

Phylum Coelenterata
Class Hydrozoa
Order Siphonophora

With its conspicuous, blue, gas-filled float, the bluebottle is one of Australia's well recognised and common stingers. Another distinguishing characteristic is the bluebottle's single main fishing tentacle, which is significant as the stinging appendage (Using vinegar for *Physalia* stings page 34).

The bluebottle belongs to the genus *Physalia*. There is still some uncertainty about the correct names for the Australian species but currently *Physalia utriculus* is used for the bluebottle with its single tentacle and *Physalia* sp. for the multi-tentacled Pacific man-o'-war which is also found in Australian waters but is not common. The name *Physalia physalis* is used for the Portuguese man-o'-war, an Atlantic species with multiple long tentacles.

The bluebottle is a floating colony consisting of feeding and reproductive polyps supported by a float that is from a few centimetres to 10 cm long and the tentacle (a protective polyp) may reach 2–3 metres (see page 30).

Large Pacific men-o'-war will have up to 7 or 8 main tentacles which may stretch to ten metres or contract to a few centimetres in length. Each tentacle has a frilled and beaded appearance due to the groups of stinging capsules, or nematocysts, spaced closely along its length. Each bead consists of hundreds of nematocysts so that the whole tentacle carries a powerful armament. The float, which is up to 15 cm long, allows the colony to sail the seas, trailing long feeding tentacles and polyps below.

The more dangerous Atlantic Portuguese man-o'-war has floats to 30 cm long and a correspondingly massive armament of tentacles up to 30 metres long. Fatalities have followed severe stings from this species (Williamson *et al.* 1996).

Bluebottles and men-o'-war generally feed on small, open-sea fishes.

Clay Bryce

◀ The bluebottle, *Physalia utriculus,* has a gas filled float which supports the single long main tentacle and feeding and reproductive polyps below.

As the venom from the the nematocysts paralyses a fish, the tentacles contract, drawing the prey up within reach of the feeding polyps on the underside of the float.

Fleets of bluebottles are often carried by warm currents to more temperate waters where, in summer, they are washed ashore on the east coast of Australia in thousands by storms or onshore winds. In south western Australia they are less abundant and are generally seen on beaches in autumn and winter, less often in summer. To date the Australian occurence of the Pacific man-o'-war has been recorded from only north Queensland.

THE STING

When the nematocysts of *Physalia* spp. are triggered to explode, their tubes uncoil with enough force to sting through a surgical glove. Each nematocyst tube may be ten to twenty times as long as the capsule (i.e. 200 to 300 microns) and carries venom from the capsule into the person.

Clay Bryce

▲ Dried bluebottle tentacles can still sting if rehydrated.

The effect of the bluebottle venom on a person is determined by the shape of the tentacle on contact. Usually, the weal is one, long, single line. The very sticky, strong, single tentacle can simply lengthen from the first point of contact and wrap around the person. The linear weal has patches where batteries of nematocysts cause particularly strong adhesion. However if the contact occurs when the tentacle is contracted to a spiral ribbon, the weals can be irregular in shape and the intensity of stinging very severe. This can happen when a whole animal is picked up — perhaps by a curious child.

Usually the sting from a bluebottle causes only skin and regional lymph gland pain with no general symptoms in adults. More severe toxic symptoms with back pain, breathing problems and anxiety have been associated with stings from the Pacific men-o'-war (Fenner *et al.* 1993).

The potent venom of the Atlantic species, the Portuguese man-o'-war, is a neurotoxin, which has been found to paralyse and kill by acting first on the nervous system, particularly the respiratory centre, and then on the muscles. It also destroys red blood corpuscles, causes bleeding and kills skin cells.

INJURY PREVENTION

- Avoid swimming when bluebottles are washed up on the beach as they are likely to be in the water as well.
- If swimming is unavoidable, protective clothing may help.
- Do not touch the tentacles of dead bluebottles on the beach — they are venomous long after the animal has died (see page 32).

SYMPTOMS

Allergic reactions are rare but may be rapid and severe, usually in a person who has been stung previously (see pages 219–223).

- Thick, raised, more or less linear, pale weals with beaded appearance, surrounded by red flare and even blistering.
- Pain, sometimes intense.
- Pain in lymph glands; in the groin if the legs are stung and in the armpit if the arms are stung.
- General symptoms such as nausea, vomiting and tender lymph glands may follow a severe sting to a child.

- Usually no general symptoms in adults, but multiple stings with a large dose of venom can produce kidney and muscle pain, general weakness and distress, and breathing difficulty.

TREATMENT

Treatment must be directed at deactivating nematocysts to prevent further stinging which would result in continuing pain. Assume that there are more unexploded nematocysts sticking to the skin, whether as an obvious tentacle or as microscopic individual cysts.

- Hot water immersion at 45°C for 20 minutes gives effective pain relief.
- In cases of general reaction with severe pain, shock and allergic reactions, seek immediate medical treatment.
- For bluebottle stings, douse liberally with vinegar for over 30 seconds (see below).
- Never use alcohol (as in methylated spirits or concentrated drinks such vodka, whisky, etc.) or fresh water as these cause immediate discharge of nematocysts.
- If tentacle is still sticking to the skin, pull off without touching directly, and pull in the line of the tentacle. It can be wound around a knife or stick, similar to winding string.
- Wrap solid ice or frozen gel with layer of towel (or cooling could be too severe) and apply. Crushed ice in a plastic bag and cool cans can be safely applied to the stung area. Assess pain at 15 minute intervals and reapply cold pack if necessary.

USING VINEGAR FOR *Physalia* STINGS

- Using vinegar to deactivate nematocysts on the skin is no longer a recommended 'all-cure' for *Physalia* stings as it causes nematocysts (stinging capsules) of the multi-tentacled species such as the Pacific man-o'-war to discharge. But vinegar does prevent discharge of nematocysts from the single-tentacled bluebottle, and so the current recommendation is: **Use vinegar for bluebottles only.**

FIRE CORAL

Millepora spp.

Phylum Coelenterata
Class Hydrozoa
Order Milleporina

Although related to hydroids and the Portuguese man-o'-war, the fire or stinging coral *Millepora* looks very different as it has a hard coral-like skeleton. Unlike the true corals, however, the surface is very smooth, with minute pores for feeding polyps, surrounded by smaller pores for stinging polyps (see page 36). The sexual stage occurs in tiny medusae borne in attached cup-shaped depressions.

In Western Australia, the most common fire coral, found on coral reefs off the north western coast, is *Millepora platyphylla*. It forms massive, stony, plate-like colonies which are yellow brown in colour with paler edges. There are also several less common, bushy species with flattened branches found in coral lagoons (see page 37). All have similar stinging abilities.

Like most marine venoms the venom of the fire coral is a protein substance and is destroyed by heat. It can cause rupturing of red blood corpuscles and may cause local skin death.

Clay Bryce

▲ The fire coral *Millepora platyphylla* is common on coral reefs in north western Australia.

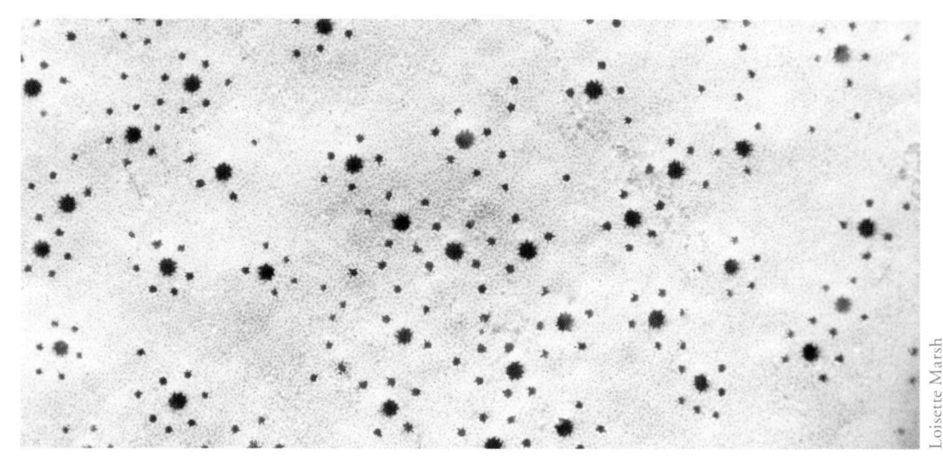

Loisette Marsh

▲ *Millepora* sp. Magnified surface of fire coral with larger pores for feeding polyps surrounded by minute pores for stinging polyps.

INJURY PREVENTION

- Wear gloves and protective clothing when diving or snorkelling on coral reefs.
- *Millepora* may be mistaken for harmless *Porites* coral so look carefully before touching or, preferably, avoid touching any coral.
- If contact is made with bare hands, keep hands away from areas with delicate skin such as eyes, lips etc.

SYMPTOMS

- Pain varies from a mild prickling sensation to a severe stinging pain followed by a burning itch with redness, swelling and weals or sometimes blisters.
- On rare occasions, a burn-like necrosis (death) of the skin follows blistering.
- Prolonged sweating over the area may follow.
- Symptoms may last for more than 24 hours but usually the pain subsides after a few hours.
- Itching often follows as the pain subsides.
- Nausea and vomiting rarely occur.

TREATMENT

Refer to treating mild to moderate stings, pages 23–24.

Seek medical attention for deep burn-like reactions.

CASE REPORT

In 1836 at the Cocos (Keeling) Islands, Charles Darwin, although not the first to note the stinging power of *Millepora*, gave a careful firsthand description of its sting. He found that the degree of stinging varied between specimens of the same species; a piece rubbed on the tender skin of the face or arm gave a pricking sensation which came on after an interval of about a second and lasted only for a few minutes but on another day 'by merely touching my face with one of the branches pain was instantaneously caused; it increased as usual after a few seconds, and remaining sharp for some minutes, was perceptible for half an hour afterwards. The sensation was as bad as that from a nettle but more like that caused by the *Physalia* ... Little red spots were produced on the tender skin of the arm but did not progress to blisters'.

(Charles Darwin, *The Voyage of HMS Beagle Round the World*, originally published in 1845)

Clay Bryce

▲ *Millepora* sp. in a coral lagoon.

Jellyfish

Phylum Coelenterata

Classes Cubozoa and Scyphozoa

The coelenterates Cubozoa and Scyphozoa have an obvious medusa stage which is why they are commonly known as jellyfish; the polyp stage being inconspicuous. Jellyfish have a central mouth, and most of the medusa bell consists of a thick jelly layer between the outer and inner cell layers around the stomach. The medusa is roughly cube-shaped in the Cubozoa (box jellyfish) and umbrella- or bell-shaped in the Scyphozoa (round or saucer jellyfish).

(Small round jellyfish which are the sexual stage of hydroid polyps are described on pages 27–29.)

Most jellyfish drift with the currents, often accumulating in large numbers in sheltered bays and estuaries. They swim with a pumping action of their bell, but for most species this does little more than allow them to move up or down in the water. The four-sided box jellyfish, however, are powerful swimmers with good control of speed and direction.

Jellyfish feed in various ways, depending on the number and length of

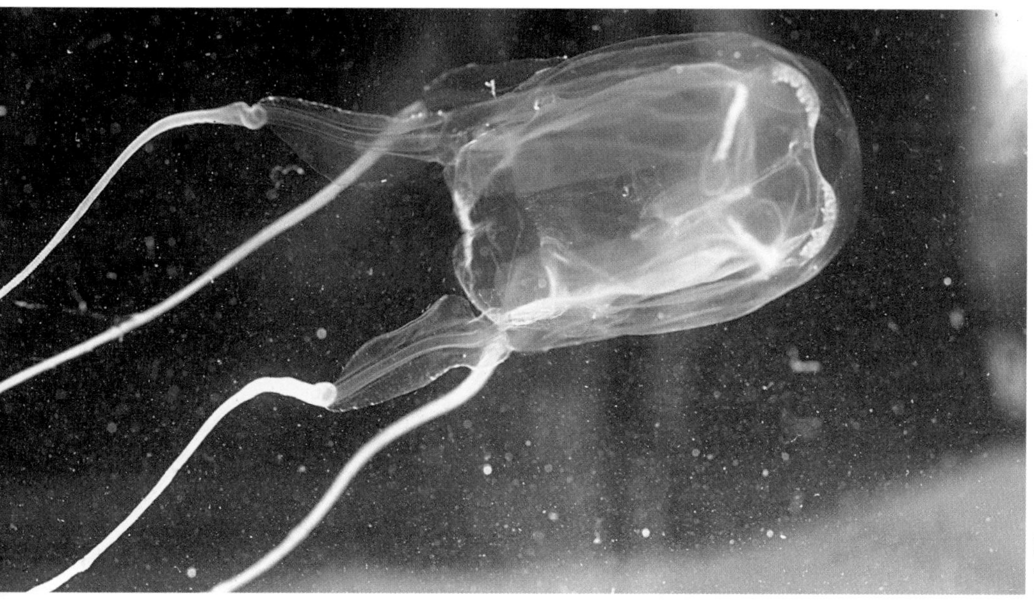

Isobel Bennett

▲ The small box jellyfish *Carybdea rastonii* is often troublesome to swimmers in sheltered sandy bays on the east and south coasts of Australia.

their tentacles and the power of their stinging capsules, or nematocysts. Some take in minute planktonic organisms and some are like corals with algae in their tissues that produce food by photosynthesis, while others catch small crustaceans and fishes with powerful nematocysts on their tentacles.

As the waters of north-western Australia become better known, more species of Indo-Pacific jellyfish are being found, and some are severe stingers. Specific identifications are not yet available for some of them. Most occur at intervals, sometimes of many years, and may appear in swarms near the coast. Some species are encountered repeatedly in the same area; for example, some box jellyfish (cubozoans) have always been found near the coast while others are usually found off shore. The highly dangerous box jellyfish *Chironex fleckeri*, *Chiropsella* sp. and at least two Irukandji-type stingers have been found in north-western Australian waters.

Jellyfish stings vary from very mild to life-threatening depending on the species and other factors. First aid treatments over the past fifteen years have changed as more is known about the reaction of jellyfish nematocysts and their venom to various chemicals, and no doubt will continue to do so in the future.

BOX JELLYFISH

Phylum Coelenterata
Class Cubozoa
Order Cubomedusae

Box jellyfish, or Cubomedusae, have a squarish box-like bell with one or more tentacles hanging from a tough gelatinous lobe or pedalium on each corner of the bell. The lower edge of the bell is folded under to form a shelf. On each side of the bell above the margin is a sense organ consisting of a balance organ and a complex eye, with lens and retina, which looks into the bell. Inside the bell is a tubular mouth, hanging from the centre, and gonads which vary in shape and position depending on the species (Figure 4). Box jellyfish are often colourless and difficult to see in the water (see page 56).

Box jellyfish are divided into three families, the Carybdeidae and

FIGURE 4:

A four-tentacled box jellyfish, or Cubomedusa, with squarish bell and each tentacle arising from a leaf like pedalium.

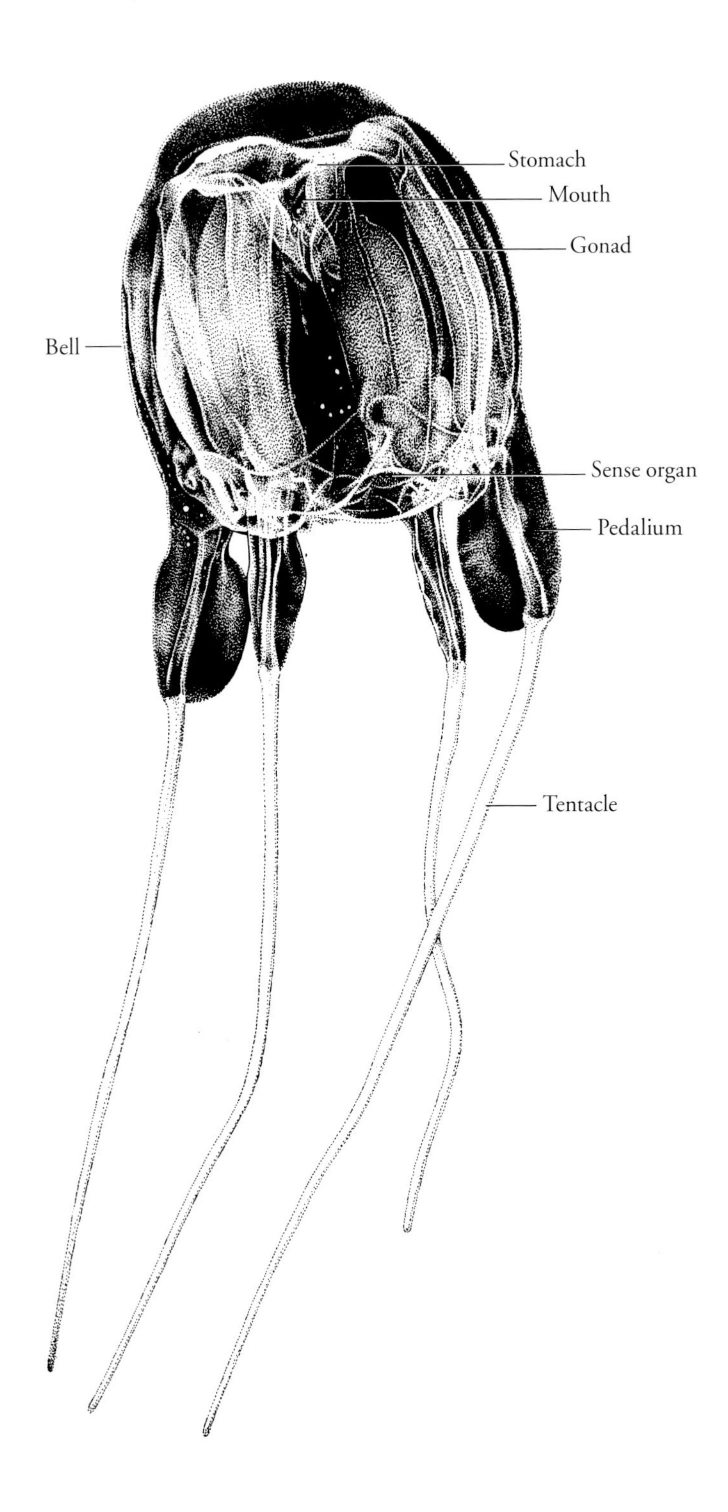

Alatinidae with four tentacles and the Chirodropidae with four clusters of tentacles. Those with many tentacles are the most dangerous.

The family Chirodropidae includes what is regarded as the most venomous animal in the world, *Chironex fleckeri* (see pages 55–65), whose severe sting can kill in as little as three minutes. Other species of Cubomedusa can be life-threatening and several are severe stingers. Most occur in tropical waters off northern Australia including north-western Australia and the Indo-West Pacific Region. First aid treatment for severe box jellyfish stings requires immediate dousing with vinegar, cardiopulmonary resuscitation (CPR) if necessary and medical treatment as quickly as possible.

Two species of Carybdeidae, relatively minor stingers, occur in southern Australia.

Small four-tentacled box jellyfish

Phylum Coelenterata

Carybdea rastonii Haacke, 1886

Class Cubozoa
Order Cubomedusae
Family Carybdeidae

The small box jellyfish *Carybdea rastonii* is sometimes called a jimble (and incorrectly, 'bluebottle'). It has a bell up to 4 cm high and 2.5–3 cm wide, with tentacles 10–15 cm long. The bell is almost transparent, and the shadows of *C. rastonii* on the sand are often more obvious than the animals themselves. The four tentacles may be cream, lavender or pink, and are more easily seen than the bell. The stomach is shallow with gastric cirri (filaments with nematocysts and glands that produce digestive enzymes) in a line across each corner.

Species of *Carybdea* frequent quiet bays, particularly those with white sand where the jellyfish collect close to the seabed when the light is bright in the middle of the day. On cloudy days and/or early or late in the day, and presumably at night they disperse through the water or swim near the surface. They feed on very small plaktonic crustaceans and larval fishes.

The early life history of *Carybdea rastonii* has been studied in Japan where reproduction takes place in summer. After internal fertilization (the sperm having been sucked into the bell from the plankton), the eggs are held in the gastric cavity of the female until they grow into swimming larvae. These emerge from the female's mouth and swim at the surface for a few days before sinking to the seabed where they settle on a hard surface and develop into small (2 mm tall) polyps with four tentacles. A single jellyfish develops from each polyp but the polyp may bud off more polyps, each developing into a medusa. *C. rastonii* apparently overwinters as a polyp, and juveniles appear in early summer.

C. rastonii is common off south eastern Australia, extending west to Albany, Western Australia, and is most common in summer and autumn. It is also found in Japan, the Philippines, Hawaii and other parts of the Pacific. On the west coast of Western Australia from Geographe Bay to Geraldton, *C. rastonii* is not present but a similar species, *C. xaymacana*, is found.

Carybdea xaymacana Conant, 1897

Class Cubozoa
Order Cubomedusae
Family Carybdeidae

Lisa-Anne Gershwin

▲ The west coast stinger *Carybdea xaymacana*.

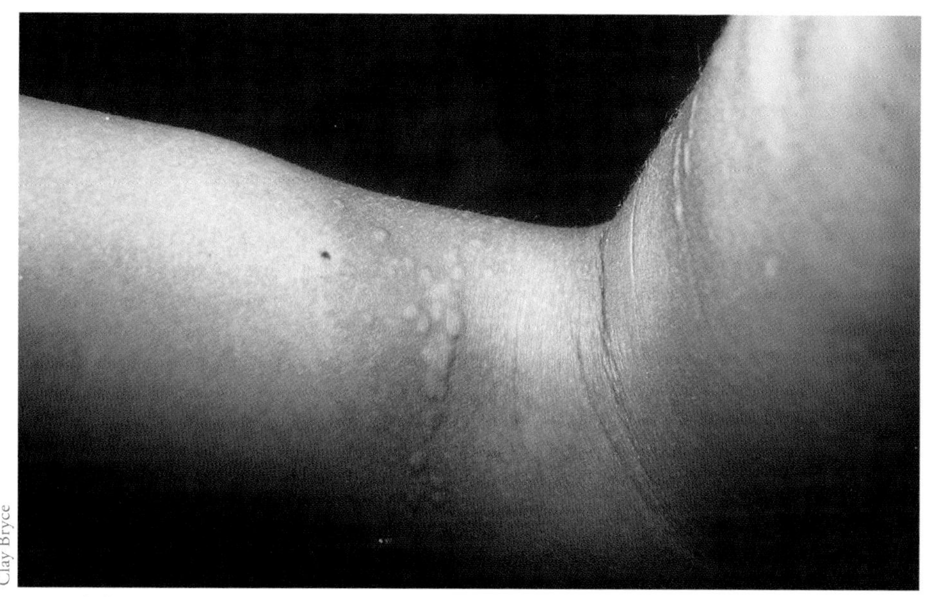
Clay Bryce

▲ Weals from a *Carybdea* sting.

The west coast stinger *Carybdea xaymacana* has a smaller bell than *C. rastonii*. The bell is usually less than 3 cm high and is transparent, with a brush-like bundle of gastric cirri in each corner of the stomach. The four tentacles are dark purple and usually less than 20 cm long.

Like *C. rastonii*, this species frequents quiet sandy bays and often swims near the surface, even on bright sunny days.

Carybdea xaymacana was originally described from the Caribbean but is now known to be more widespread in its distribution (Gershwin, pers. comm.). In Western Australia it is abundant in Geographe Bay, off Perth beaches and off Rottnest Island and Geraldton. It has also been found in north Queensland off Cairns and Cooktown.

INJURY PREVENTION

- When there are reports of stingers, usually from other swimmers, wear a cotton or lycra top with long sleeves and a high neck while swimming; stings on the legs with their thicker skin are not usually a problem.
- Be more wary in mid to late summer when the jellyfish are larger and more prevalent.

SYMPTOMS

- Sharp stinging sensation on sensitive skin, which may increase or lessen within five minutes, followed by smarting or itching of the area.

- Severity of the sting will vary with the sensitivity of the skin. A sting on the outer side of an arm may be mild but on the armpit or lips or inside a swimsuit, can be very painful.
- Initially up to four red lines from the four tentacles may mark the area of contact. Redness spreads and becomes blotchy and 1–4 weals 0.5–1 cm across may develop.
- Effects usually disappear in an hour or two though brown lines may remain for several weeks. A severe sting may be painful for 12 hours or more with the weal persisting and ulcerating over the next few weeks and with the brown colour lasting many weeks or months.

TREATMENT

- Do not use alcohol (as in methylated spirits or concentrated drinks such as whisky, vodka, etc.).
- Wash with vinegar to deactivate nematocysts and prevent further stinging.
- Soak stung area in water as hot as can be borne (45°C) without further injury or scalding; use an unstung area to test temperature. As water cools, replace with more hot water because pain will not ease unless treatment is continued for at least 20 minutes (Taylor, 2000, 2007).
- Do not use fresh water unless it is hot.
- Never rub or wash the skin vigorously, which could spread the nematocysts and, by applying pressure, cause nematocysts to discharge.

Refer to treating mild to moderate coelenterate stings, pages 23–24.

IRUKANDJI SYNDROME

The Irukandji syndrome was named in 1952 after an aboriginal tribe living in the Cairns area of north Queensland for the group of reactions resulting from a sting by a very small box jellyfish. This box jellyfish was later named as *Carukia barnesi* after its original investigator, Dr Barnes, who had identified the cause of Irukandji syndrome by experimenting on himself, his son and a friend, with all three landing up in the intensive care ward of the Cairns Base Hospital (Southcott 1967).

Carukia barnesi has a thimble sized bell up to 1.0 cm high by 1 cm wide, with four tentacles varying from 4 cm to 1 m in length depending on the degree of contraction. Stinging capsules are grouped in minute red warts on the body and tentacles (see page 45), with nematocysts on the bell differing from those on the tentacles.

In recent years, variations in this type of syndrome — with two fatalities and a number of life-threatening cases occuring in north Queensland and north Western Australia — have suggested that more than one species might be involved. Efforts to catch and identify the culprits have found at least nine other species of small box jellyfish thought to be responsible. These have been collected but are not *yet all* identified, and some may be undescribed. Two species of small carybdeids, recently described, have been found off Broome, Western Australia. The Irukandji syndrome can also be triggered occasionally by other species such as *Carybdea* spp., *Alatina* sp. and *Physalia* spp.

In Western Australia, stings resulting in the Irukandji syndrome

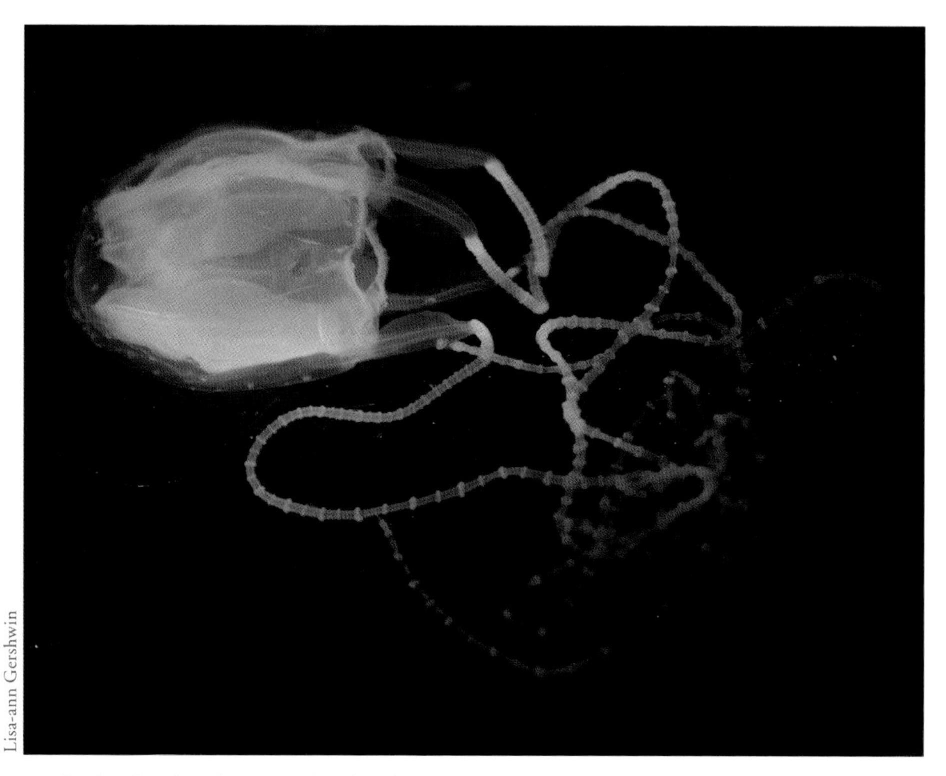
Lisa-ann Gershwin

▲ Closely related to the original Irukandji stinger, *Carukia barnesi*, a newly described species, *Carukia shinju*, was found in the pearling grounds off 80 Mile Beach, Western Australia. The species name '*Shinju*' means 'pearl' in Japanese. The bell is about 16 mm in height.

have been reported from off the Rowley Shoals, Broome, Onslow and Exmouth. All of these stingings were from small, unidentified, carybdeid jellyfish other than in the Exmouth case which was from a large species of *Alatina*. *Malo maxima*, with a 4 cm high bell has caused cases of severe life threatening Irukandji stinging.

Carukia barnesi has not yet been identified from Western Australian waters although*Carukia shinju* Gershwin, 2005 with similar symptoms was collected off 80 Mile Beach.

Unlike *Chironex fleckeri* which spends its whole life cycle near the coast, particularly in muddy waters, so-called 'Irukandji stingers' appear to prefer clear water in the open ocean and so may be encountered by divers at remote coral reefs. However, they may be carried inshore by winds and currents, and stings triggering the Irukandji syndrome have occurred at Cable Beach, Broome.

The danger area is from North West Cape, Western Australia, across northern Australia to south of Fraser Island, Queensland. Most stings occur in the afternoon on fine days when the jellyfish swim near the surface. Although these jellyfish are most numerous in North Queensland in January, the danger season in north Queensland is from September to May. However, the incidence of Irukandji stingings is very variable; in some years there will be hardly any cases while in other years, 100–200 cases of stingings will be reported. The severity of the sting also varies.

At Broome, Western Australia, stings are most likely from November to May but Irukandji stings have also occurred during other months.

In Broome in 2003 from December to March 90% of stings occurred in the east-facing Roebuck Bay while in March-June 95% of stings occurred at west-facing Cable Beach (Macrokanis *et al* 2004).

INJURY PREVENTION

- Avoid swimming when jellyfish and other gelatinous planktonic organisms are brought inshore with clear oceanic water.
- Wear protective clothing when swimming during the stinger season in north-western and northern Australia.
- When diving wear a wetsuit and protect the neck from stings by wearing a hood.

SYMPTOMS OF IRUKANDJI SYNDROME

The initial symptoms are mild and last about half an hour. General symptoms follow, usually 30 minutes after the sting but could be up to two hours later. These symptoms can be severe — Irukandji syndrome can be life-threatening. With the easing of initial symptoms, the victim might be tempted to go back into the water and then may develop further symptoms there, which would increase the danger. Suspected victims should not re-enter water for at least two hours in case general symptoms develop. In divers, Irukandji symptoms may be confused with decompression illness.

INITIAL SYMPTOMS

- Sensation on contact varies from a sharp stinging pain 'like a bee sting' (Fenner 1999) to a vague underwater touch with the victim possibly unaware of any serious injury.
- Sometimes, within five minutes of contact, area around the sting develops a mild blotchy redness and small red 'goosepimple' lumps appear, increasing for 20 minutes, before subsiding.
- Red coloration can last up to three hours, and skin is dry at first and then sweats profusely over stung area.

GENERAL SYMPTOMS

I) INITIAL GENERAL SYMPTOMS ARE MUSCULAR:

- General symptoms are usually felt after about half an hour but the interval can be anywhere between five minutes and two hours;
- Development of severe lower back and abdominal pains with rigidity of the abdominal wall; muscular aches, cramps in all four limbs, joint pains and backache, and muscular tenderness, which may be severe (abdominal pain may be confused with an internal illness and chest pains with heart attack); and unbearable pain coming in waves of increasing intensity without any intervals of freedom from pain.

II) FURTHER SYPTOMS:

- increased sweating, pallor, anxiety and restlessness, with a sense of 'impending doom';
- severe headache;
- fast and often irregular heartbeat;

- temperature usually remains normal;
- incessant nausea and vomiting,
- in *Carukia* spp. the main symptoms diminish or cease in 4 to12 hours with some ill effects persisting for several days;

SEVERE IRUKANDJI SYNDROME (AS FROM *MALO MAXIMA* AND OTHER SPECIES):

- dangerously high blood pressure (up to 300/180mmHg) which can produce risk of bleeding in the brain and stroke as in both of the deaths (in Queensland).
- respiratory distress and coughing due to pulmonary oedema (fluid in the lungs) which could indicate heart failure.
- Recovery is usually complete but can take weeks.
- No scarring from sting.

TREATMENT

- Do not use alcohol (as in methylated spirits or concentrated drinks such as whisky, vodka, etc.) or fresh water.
- Douse stung area with vinegar to deactivate nematocysts on skin (this will have no effect on pain resulting from venom already discharged into body).
- Do not allow victim to return to water.
- Use nitrate spray under the tongue to relieve high blood pressure (Fenner and Lewin 2003).
- Get to hospital or if not possible, send for medical aid. The victim should stay under medical care until it is evident that full syndrome will not develop. Some victims will require intensive care which may include control of blood pressure, intravenous pain relief and treatment of heart failure.
- Keep victim still to limit the spreading of the venom.
- Clear water of swimmers, particularly when sting has occurred in netted swimming area.

MEDICAL MANAGEMENT

- There is no antivenom available yet.
- Intravenous magnesium sulphate is very effective in reducing all symptoms including pain and high blood pressure (Corkeron 2003).

CASE REPORTS

In April 2001, a jellyfish sting suffered by a young woman swimming at Cable Beach, Broome, Western Australia, caused cardiac arrest. She was treated in intensive care in Perth and spent more than a week in hospital. While the responsible jellyfish was not caught, specimens which local pearl divers believed to be Irukandji proved to belong to an undescribed species.

(The West Australian, 3 May 2001)

In 1998 a report of 60 Irukandji stings near Cairns showed no life-threatening effects

(Fenner 2005).

In 2002, the Cairns Base Hospital team reported on 50 cases of Irukandji syndrome admitted, where skin scrapings identified the nematocysts of *C. barnesi* in 39.

(Huynh *et al.* 2003)

In 1997 a scuba diver in Mackay, Queensland, with Irukandji syndrome actually described and drew the jellyfish which had touched his elbow after the treating doctors had realized his condition was not from Decompression Sickness ('the bends'). This was the first time the animal had been seen, described and drawn by the victim.

(Hadok 1997)

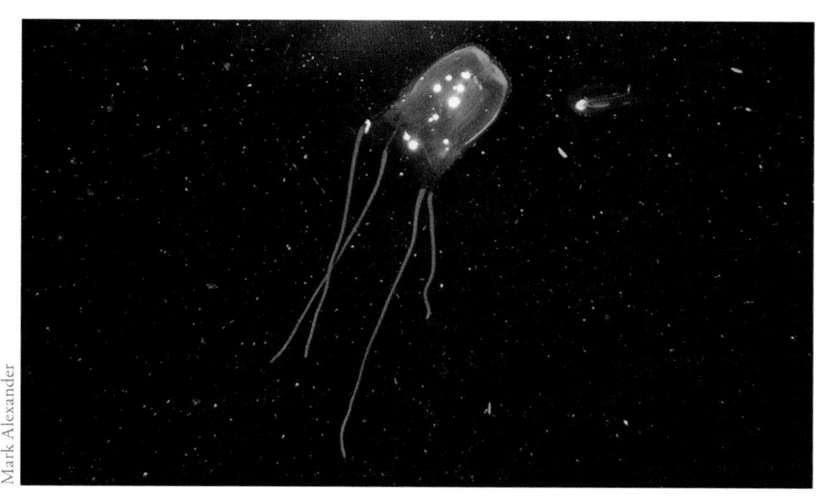

Mark Alexander

▲ This Irukandji stinger, *Malo maxima* Gershwin, 2005, has been responsible for several life-threatening cases in Broome. The bell is about 4 cm high.

Large four-tentacled box jellyfish

Phylum Coelenterata

Alatina spp.

Class Cubozoa

Order Cubomedusae

Family Alatinidae

Four-tentacled box jellyfish *Alatina* spp. can grow to a large size with a bell 40 cm high and about half that in width. Nematocyst warts may or may not be present on the bell. The four tentacles are each borne on a large lobe or pedalium which has thin flaps on either side of a stiff core. A canal running through the core leads to the tentacle. (Specimens with the bell 18 cm high had pedalia 5 cm long by 2 cm wide.) The stomach is shallow with a crescent-shaped area of short gastric cirri in each of the four corners. The mouth is at the end of a very short tube called the manubrium, hanging in the centre of the bell. On each of the four sides of the bell above the lower edge is a dark brown eye spot which is part of a sense organ in a pit enclosed by three flaps, one above and two below (Figure 5). The pit also houses a balance organ.

Several large *Alatina* specimens have been trawled in Western Australian waters north of Port Hedland and one was collected at the surface near Scott Reef, north of Broome. *Alatina* spp. are widespread

FIGURE 5:

Alatina sp. (from a photograph in Halstead, 1965)

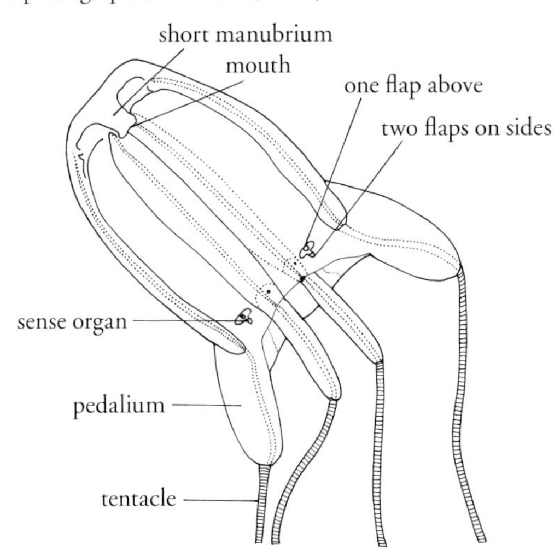

in the tropical Indo Pacific region.

Very large specimens of *Alatina grandis* have been encountered off the North West Cape peninsula. These were originally identified as species of *Tamoya* but Gershwin (pers. comm.) has shown that *Tamoya* is an Atlantic genus and is not found in Australian waters.

INJURY PREVENTION

- Wear a wetsuit when diving in tropical waters — nematocysts of *Alatina* sp. can penetrate a lycra suit.
- *Alatina* species are usually found in the open ocean so be especially observant and look above you when approaching the surface from a dive.

SYMPTOMS

Symptoms from a minor sting at Scott Reef, off north-western Western Australia, are listed below.

- Intense pain and a weal with surrounding redness were caused by less than one centimetre of tentacle.
- Tightness in chest which eased after about half an hour.
- No scarring from sting.

SYMPTOMS FROM A STING BY A LARGE *ALATINA* SP. NEAR NINGALOO REEF.

Stings from *Alatina* spp. are not life-threatening. General symptoms do not usually follow the initial symptoms but in a few cases the victim may require hospitalisation.

- Almost instant, severe, burning pain.
- Pain soon spreads to lymph glands.
- Intense pain which lasts about two hours, and then slowly subsides over several hours.
- No ulceration, skin death or scarring from sting.

TREATMENT

- Do not use alcohol (as in methylated spirits or concentrated drinks such as whisky, vodka, etc.).
- Wash with vinegar to deactivate nematocysts and prevent further stinging.

- Soak stung area in water as hot as can be borne without further injury by scalding — use an unstung area to test temperature. As water cools, replace with more hot water because pain will not ease unless treatment is continued for at least 20 minutes (Taylor 2000).
- Never use fresh water unless it is hot.
- Never rub or wash the skin, which could spread the nematocysts, or apply pressure thereby causing nematocysts to discharge.

Refer to treating mild to moderate coelenterate stings, pages 23–24

CASE REPORT 1

In March 1978, a US Navy serviceman was stung on the right foot in knee-deep water near Exmouth, Western Australia. There was immediate pain, which increased. After five minutes, a slight 'knot' in the lower abdomen progressed to just below the rib cage and began to interfere with breathing. Abdominal and chest muscles became rigid and breathing distressed. Although the pain was primarily in the foot, it spread to the base of the spine and then to the whole body, including the head. Loss of motor control below the neck followed together with cold, clammy skin and profuse perspiration. Administration of oxygen did little to assist breathing.

In hospital, he was treated with intravenous cortisone and painkillers every hour for nearly 24 hours. The pain gradually retreated to both feet and he was discharged from hospital two and a half days after admission but was unable to work for a further three days.

Three weeks later, 89 small red spots were counted in four areas of the instep. These disappeared over the following week with no scarring by which time his strength had returned. He lost three kilos in weight during his ordeal.

The morning after the stinging, six dead jellyfish with bells 25 cm in length were found. They were photographed and identified from the photographs as *Tamoya* sp. by the late Dr R. Endean of Queensland University, who regarded the symptoms of the serviceman as an Irukandji syndrome reaction (see pages 44–48).

Endean, unpublished, and photograph taken at Exmouth, WA (p86, Edmonds 1989)

Patrick Baker

▲ *Alatina* sp. photographed off Ningaloo Reef, Western Australia. The bell was about 40 cm long with orange nematocyst warts on the surface, and the tentacles were completely contracted.

Geoff Taylor

▲ *Alatina* sp. Exmouth.

CASE REPORT 2

A photographer from the Western Australian Museum was scuba diving outside Ningaloo Reef in March 1987 when he saw several very large carybdeid jellyfish about 40 cm long, with very short tentacles and orange warts on the surface of the bell (see page 53). He cautiously touched one with the fingers and palm of his right hand. About a minute later the palm and fingers began to tingle, then hurt more and more, with 'all the nerve ends crying out!' A dull pain developed under the left arm, followed by tightened breathing. He was taken some distance back to camp, and Caladryl™ lotion was applied. The pain gradually subsided after several hours rest and no further general symptoms developed.

(Patrick Baker, unpublished)

CASE REPORTS 3 AND 4

In 2000, Dr G Taylor reported in the South Pacific Underwater Medicine Society (SPUMS) Journal:

'The next two cases are my own experiences of being stung on two occasions and the successful use of heat to treat the pain of envenomation:

The first occasion occurred on Ningaloo reef in 1994. While swimming in deep water awaiting pick-up by a boat, my left knee struck a sizeable *Alatina*. The sting penetrated through a lycra bodysuit, causing instantaneous severe burning pain. The pain soon spread to regional lymph glands in the groin, but there were no systemic symptoms. The intense pain lasted for about two hours, and then slowly subsided over the ensuing three hours.

The second event occurred in the same locality, a year later. On this occasion, while snorkelling, my head struck the *Alatina*, the stingers penetrating my hair (which is surprisingly thick), with extensive stinging over the scalp. At the same time I lifted my hand in a reflex action to fend off the 'attacker', and was stung on the back of the hand.

On this occasion it was decided, as an experiment, to try treating the sting with heat. My hand was immersed in a bowl of hot water as hot as I could stand. This brought almost immediate relief of the pain, but initially the pain recurred after removal from the heat. After 20 minutes the effect of the heat treatment was persisting. However, the pain in the scalp had spread to the neck probably through lymphatic spread. There was an intense burning, and a feeling as if my head was in a vice.

Hot towels were tried and a hot shower, but in the end I was subjected to lying prone on a bench with my head in a bowl of hot water. The relief was very rapid and after 20 minutes the pain had lessened to such a degree that treatment was ceased. Within an hour of being stung I was virtually pain free, and able to resume diving.'

Large box jellyfish with many tentacles

Phylum Coelenterata

Chironex fleckeri Southcott, 1956

Class Cubozoa

Order Cubomedusae

Family Chirodropidae

The large box jellyfish with many tentacles, *Chironex fleckeri,* is the most dangerous jellyfish in the world. Since first documented in 1884, 64 people have met sudden and painful deaths in tropical Australian

Roger Steene

▲ A young specimen believed to be of the deadly box jellyfish, *Chironex fleckeri*, in Sabah, Borneo.

waters, 36 in Queensland and 27 in the Northern Territory. Many of the fatalities are young children in remote areas who died within minutes of being stung. Records of non-fatal stingings show 93 in Queensland, 172 in the Northern Territory and 5 in Western Australia (Williamson *et al.* 1996). More injuries and deaths attributed to *C. fleckeri* have been recorded in Indo-Pacific countries north of Australia.

Chironex is the largest of the box jellyfish with many tentacles. Although the squarish domed-topped bell may reach 30 cm across and weigh more than 6 kg, adults are more commonly 10–17 cm in bell height. They have up to fifteen ribbon-like tentacles arising from each of the four claw-like pedalia, and each tentacle can grow to a length of 2 m. Smaller specimens of *Chironex* have correspondingly fewer and shorter tentacles.

The largest specimen of *Chironex* reported carried a total of more than 91 metres of tentacles. *Chironex fleckeri* was formerly confused with a related box jellyfish *Chiropsella* sp., a less dangerous stinger. The two species can be distinguished at any size by the shape of the canals running through the pedalia to the tentacles. In *Chironex* each canal bends at an acute angle projecting upwards, while in *Chiropsella* the canal has a right angled bend (arrows in Figure 6). In immature animals the gonads of

John Williamson

▲ *Chironex* is responsible for many deaths in northern Australia. From the sea surface, a swimming *Chironex* is almost invisible.

FIGURE 6:

A comparison of the diagnostic features of *Chironex* and *Chiropsella* (drawn from photographs by Barnes, 1966), as *Chiropsalmus*.

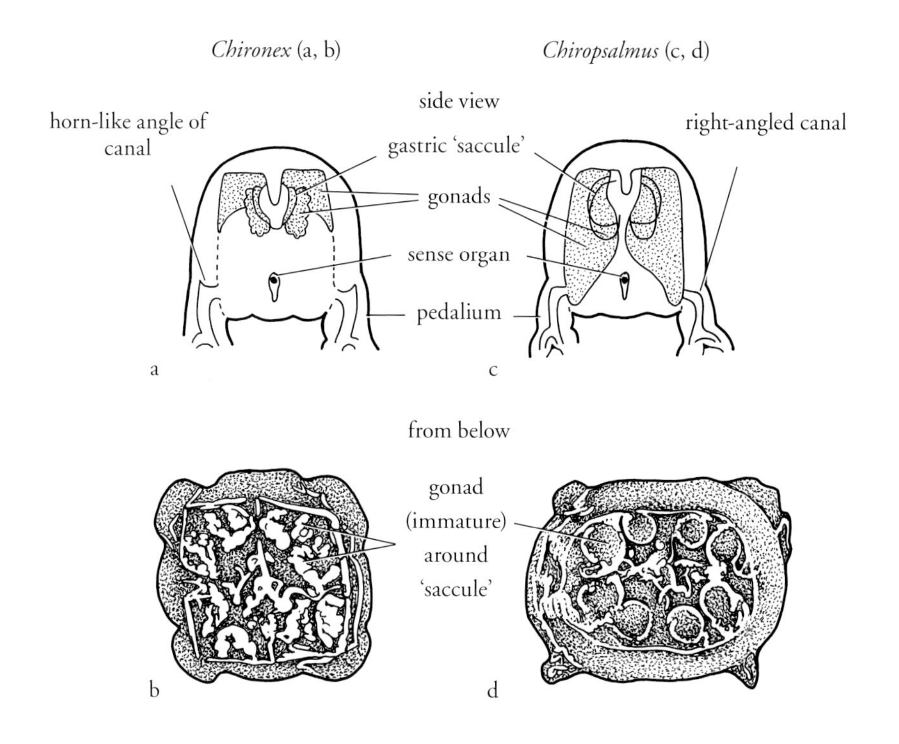

Chironex resemble cockscombs while those of *Chiropsella* are smooth and rounded. When mature, the gonads of *Chironex* hang down like bunches of grapes inside the bell (Barnes1966).

Chironex and *Chiropsella* of the same size have about the same number of tentacles but those of *Chironex* are longer and more heavily armed with nematocysts. A specimen of *Chironex* would have about ten times the volume of stinging cells of a *Chiropsella* of the same size. However, *Chironex* grows to a far greater size than *Chiropsella* and acquires almost twice the number of tentacles, each of which is longer and thicker, so that a well-grown *Chironex* could carry 100 to 200 times the amount of venom of a full sized *Chiropsella*.

Chironex are found in inshore waters of northern and eastern Australia from Point Samson, near Roebourne in north-western Australia, to the

Gladstone area in eastern Queensland. They feed on small fishes and shrimp abundant near the shore in calm summer conditions, usually in protected sandy or muddy bays. Some have home ranges of only 500 m. An undescribed species of *Chironex* has been found near Broome, WA.

Chironex can swim rapidly (up to 5 km per hour) and change direction abruptly when hunting. They remain seaward of breaking waves to avoid turbulence. During rough weather they are thought to go to the seabed in deeper water. They are able to control their position in the water by means of a balance organ (statocyst) and eyes, all located in a sense organ mid way along each side of the bell. The eyes are quite complex with lenses and pigment cells which may function as a retina. Although it is uncertain whether they can form an image, they do avoid obstacles in the water.

Adult *Chironex* are believed to release their eggs and sperm in coastal creeks, generally near mangroves, early in the dry season, usually April or May. The resulting larvae (planulae) settle on the under surface of rocks in saltwater creeks where they grow into small creeping polyps. These develop into attached polyps that bud off further polyps, which feed on plankton until October or November when each polyp metamorphoses into a single box jellyfish, 1–2 mm in diameter. The jellyfish migrate out of the creeks before the summer rains but stay in shallow water close to beaches. They spend their lives within a few kilometres of the coast so are not found in clear offshore water, where the Irukandji box jellyfish may be encountered (Hartwick 1991).

As *Chironex* grow, they eat larger and larger prey, until a full-grown *Chironex* 25 cm wide can catch, kill and digest a fish of equal size in less than three hours. To do this, its venom must be potent enough to cause paralysis or death, or both, within seconds so that its jelly body is not destroyed by a thrashing prey. Unfortunately, humans may accidentally blunder into *Chironex* and receive a deadly dose of venom.

Large adult *Chironex* usually appear in November, together with very small juveniles. The juveniles mature in three or four months so it is likely that the greatest number of mature specimens will be found in March or April. The only months in which no stingings have been reported in Queensland are June, July and August, but in the Northern Territory cases have occurred in these months as well.

Hospital centres have reported significant differences in the severity of

Chironex stings in the location within their region and within a season. Where one reported desperate cases needing intensive pain relief and intravenous antivenom, another reported surprisingly low-key symptoms and medical activity (O'Reilly *et al.* 2001).

At least five *Chironex* stingings have been recorded in north-western Australia, but no Western Australian fatalities have been linked with this species (Williamson *et al.* 1996).

Although *Chironex fleckeri* has not been identified from waters to the north of Australia, deaths from large box jellyfish stings are known from Bougainville Island, the Philippines and Borneo. *Chironex* are suspected to be the cause of these fatalities as *Chironex*-like chirodropids have been identified from these areas as well as Vietnam, Indonesia and Malaysia.

CHIRONEX STING

Chironex has different types of nematocyst. Those which deliver most toxin are cigar shaped. The butt has spirals of spines which open out to anchor the nematocyst during firing. The piercing thread rotates like a drill, releasing toxin while it is drilling and when it is at its full depth of up to 550 microns. Two of the other nematocyst types are short; one is adhesive, the other acts as a grappling hook (Rifkin in Williamson *et al.* 1996).

Different combinations of factors determine to what extent the venom will cause skin death and destruction of blood cells, and whether it will have a lethal effect on the heart.

Usually the venom is injected into the lower layer of the skin. Here the nematocyst tubules may penetrate small blood vessels so that venom is injected directly into the circulatory system and quickly carried to the heart (Williamson *et al.* 1996).

The severity of the sting depends principally on the length of the tentacle in contact relative to the size of the victim, the thickness of the skin in the contact area and the duration of contact. Death is probable if the total length of weals on an adult is greater than 6 or 7 metres (Barnes 1966) while about 2 metres of tentacle could kill a child. Children wading or playing in shallow water are the most frequent victims.

The size of the jellyfish and the time since the nematocysts were last used for feeding or defence affect the amount of venom discharged which in turn is a determining factor in the effect of the sting.

The severity and effect of the sting also relate to the pattern of contact. If the long tentacles of the feeding animal touch a human in shallow calm water, they adhere strongly and then contract. The concentrated batteries of stinging capsules immediately inject large amounts of powerful venom, meant to stun crustaceans or fishes. In a human, the venom causes local skin death (dermonecrosis) and spasm of all muscles including those of blood vessels and heart. Death may follow because the heart muscles cannot or may only be able to partly relax and so allow the heart to refill normally for the next beat. This may produce cardiac shock as in a heart attack. If the heart's left side is affected fluid accumulates in the lungs causing difficulty in breathing and even frothing at the mouth. The combination of the heart muscle and the arteries going into spasm is thought responsible for the pain, weird fluctuations of blood pressure and general condition of the victim.

The difference between life and death of the victim depends upon the dosage of venom injected and the size of the victim. Cardiac damage

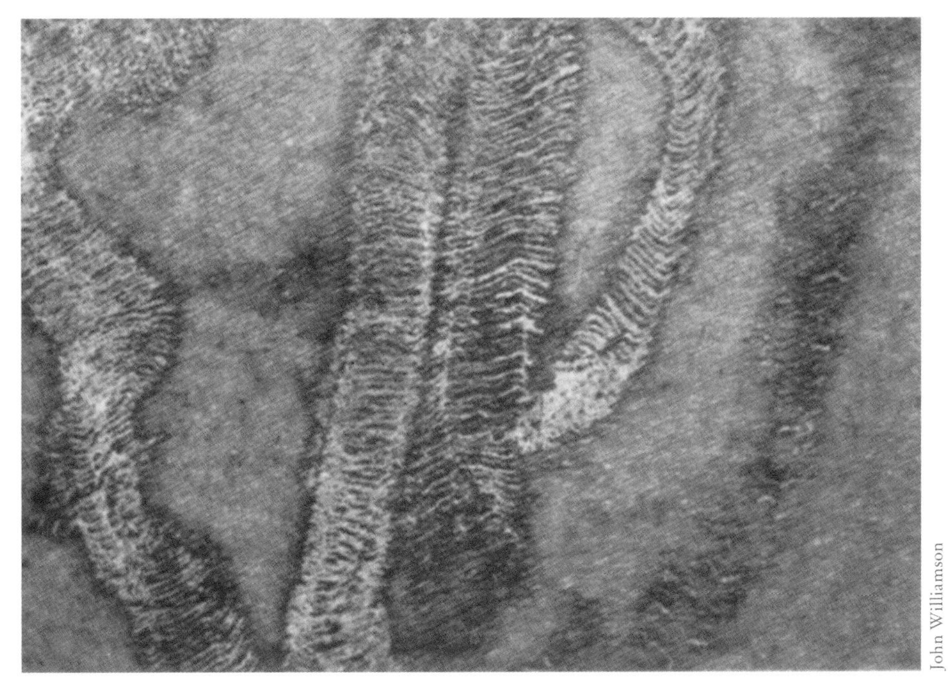

John Williamson

▲ The typical ladder-like pattern of a fresh sting by tentacles of *Chironex fleckeri*. This is followed by skin death of the area and permanent scarring.

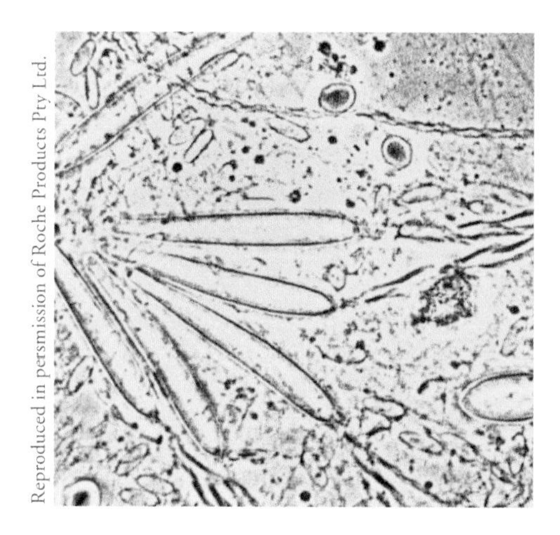

FIGURE 6A:

Highly magnified photograph of discharged cigar-shaped nematocysts of *Chronex fleckeri.*

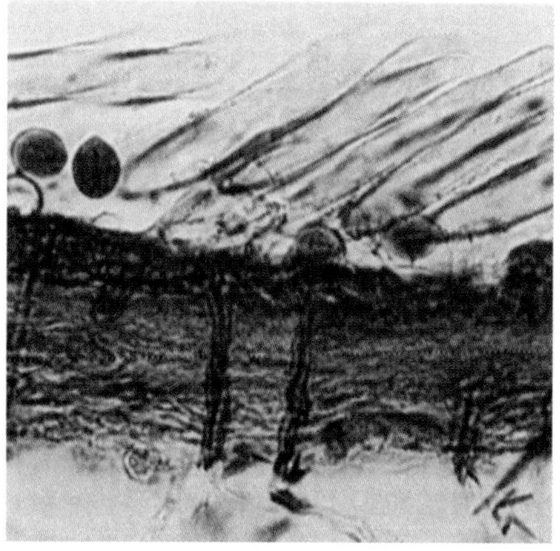

FIGURE 6B:

Magnified section through skin showing nematocysts adhering to the surface and those which have penetrated it.

tends to be an all or none response, which means that below a certain threshold amount of injected venom there is no danger to the heart but above it, there is. This is of significance in that, although rescuers may be stung by the tentacles they remove from the victim's body, they will not be in danger of losing their own lives.

The venom contains a protein which is destroyed by heating to 45°C for five minutes or by treating with a protein splitting enzyme. A traditional treatment in some places is to apply the juice of green pawpaw which contains the enzyme papain. In Queensland, *Chironex* box jellyfish antivenom has been extremely effective for pulling people out of a state of intense distress, relieving pain and for promoting the future healing of the skin. The antivenom is made from hyper-immunised sheep serum and so does not cause adverse reactions, unlike snake antivenoms which are made from horse antisera (Williamson *et al.* 1996).

INJURY PREVENTION

- June, July and August are regarded as safe months for swimming in Queensland but in the Northern Territory and the Kimberley coast, protection should be used at all times in inshore waters.
- Full length lycra stinger suits, overalls, wetsuits or clothes with long sleeves and long trousers give protection.

- Launching boats can be just as hazardous as swimming.
- Avoid entering water without protection, even to paddle in the shollows, in northern Australia.
- Avoid entering water if stingings have been reported in area, regardless of time of year or whether jellyfish have been seen.
- Regard abundance of small shrimps, especially the small red shrimp *Acetes australis*, and of small fishes in water or washed up on beach during summer as a danger sign since these are the preferred food of *Chironex* and *Chiropsella*.
- Be particularly careful near estuaries.

SYMPTOMS

- Distraught victim often staggers towards the beach and is likely to collapse. Victim may not be able to reach shore, even though in shallow water — especially children.
- Excruciating pain, with victim often becoming irrational.
- Victim may lapse into a coma and breathing may even cease.
- Central chest pain, collapse, rapid feeble and irregular pulse similar to heart attack.
- Difficulty in breathing and even frothing at mouth due to fluid in lungs.
- Fluctuations of blood pressure.
- Strands that look like 'apple jelly' or 'uncooked prawn flesh' over skin, which rapidly develop deep burn-like wounds arising very fast in a broad ladder-like pattern with permanent scarring.
- Delayed skin reaction 7 to 10 days (or more) later, painless but itchy eruption at site of original sting (which may be an allergic reaction, see pages 219–223).

TREATMENT

Act immediately as death from major sting can occur in minutes, particularly in children. First aid for Chironex stings should include immediate application of vinegar and CPR (see pages 229–231).

- Retrieve victim from the water and restrain, as any movement will increase venom absorption.
- Douse stung area and adhering tentacles liberally with vinegar for at

least 30 seconds to deactivate nematocysts and prevent further stinging. Vinegar will not relieve pain already inflicted.

- If no vinegar is available, acidic soft drinks or lemon juice may help.
- Do not use fresh water or alcohol (methylated spirits or concentrated drinks such as vodka, whisky, etc.) as these will cause further discharge of nematocysts, increasing the sting.
- Never rub with sand (or anything else), which will only increase envenomation.
- Send others for ambulance and medical help; getting victim to hospital is critical.
- Apply cardiopulmonary resuscitation (CPR) if necessary, and keep it up (see pages 229–231).
- Medical treatment for cardiac arrest, respiratory arrest, coma or severe pain should include three ampoules (20,000 units each) of antivenom intravenously.

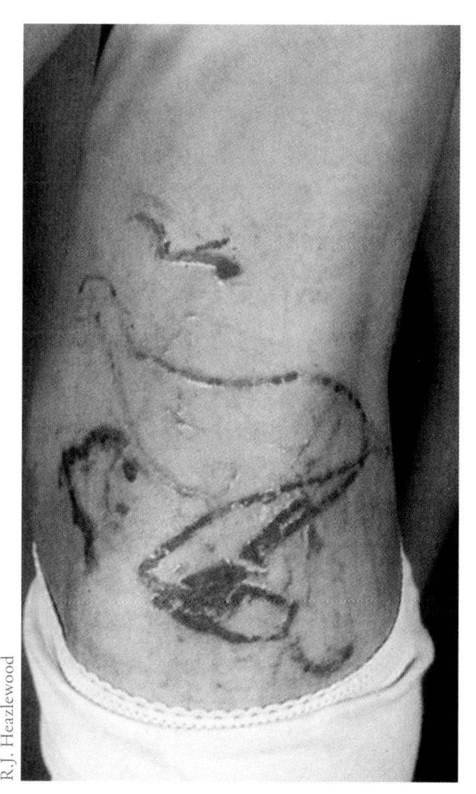

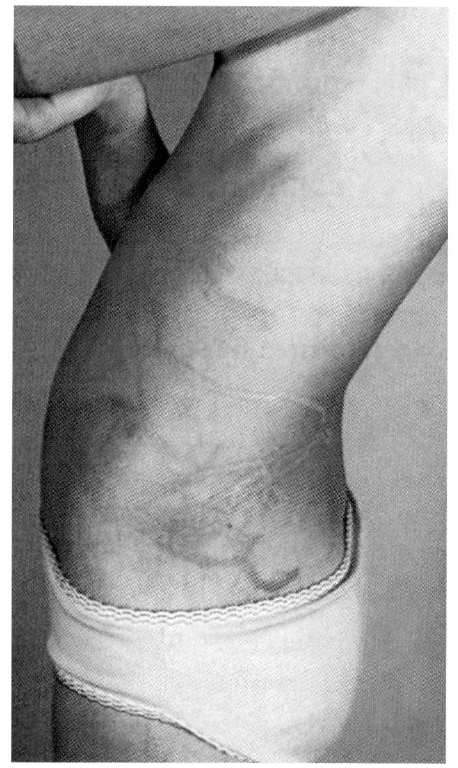

R.J. Heazlewood

▲ The stung area after one month, and after seven months.

- Remove tentacles (now harmless) sticking to skin.
- Apply ice packs, crushed ice in a plastic bag or iced drink cans wrapped in a cloth to stung area. (Effect of application of heat has not yet been reported.)
- If far from medical help, apply crushed leaves of beach morning glory, a traditional remedy for stings (see pages 22).
- Because venom is injected directly into bloodstream, external application of substances may not help apart from the immediate application of vinegar to deactivate nematocysts.

CASE REPORT 1

A child stung at Lagrange Bay, south west of Broome, in March 1975, was saved by prompt resuscitation. She showed the characteristic symptoms including the heavy scarring of *Chironex* sting, giving fairly positive identification, although the box jellyfish was not collected.

(Sutherland 1983)

CASE REPORT 2

The following case was reported in *The West Australian*, 24 March 2003:

'A boy died after being stung by a box jellyfish in North Queensland yesterday. The seven-year-old was stung across the chest and neck at Mission Beach, south of Cairns, about 1.15pm, a Queensland Ambulance Service spokeswoman said. The boy was in full cardiac arrest by the time an ambulance crew arrived with a doctor a short time later. Attempts were made to revive the boy on the way to Tully Hospital. He was pronounced dead at the hospital at 2.30pm.

'Vinegar was applied liberally and at once but cardiopulmonary resuscitation (CPR) was not attempted until the ambulance arrived.'

First aid for Chironex stings should include immediate application of vinegar and CPR.

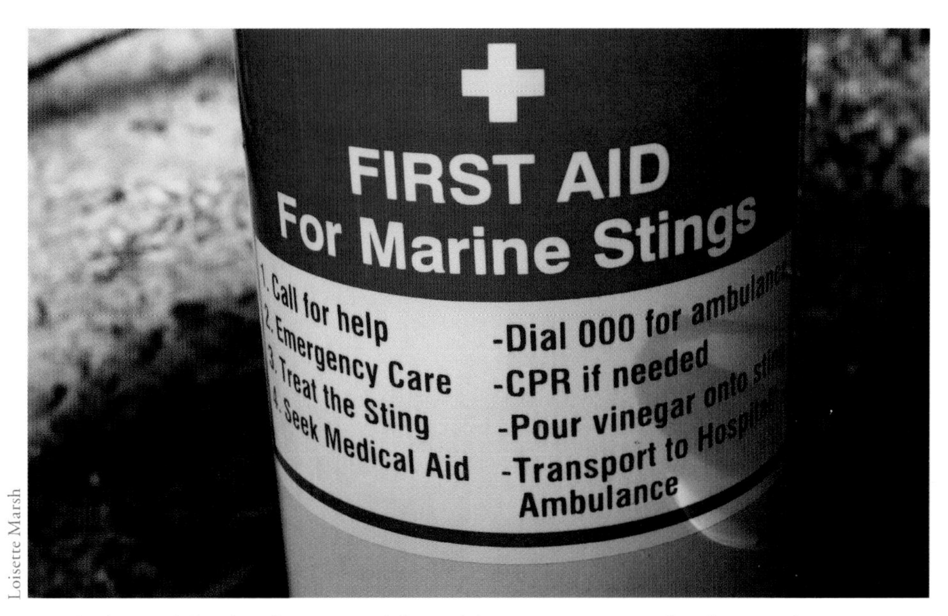

Loisette Marsh

▲ Cylinders with bottles of vinegar and first aid for severe stings are placed near each entry to a north Queensland beach.

Loisette Marsh

▲ Sign and bottles of vinegar on a north Queensland beach.

Chiropsella spp.

Class Cubozoa
Order Cubomedusae
Family Chiropsalmidae

The '*Chiropsalmus*' sp. found in north Queensland and known as 'quaddie' or 'cubo' was previously identified as *C. quadrigatus*, but is now recognised as belonging to a new genus and species. The identity of related Western Australian specimens of *Chiropsella* is still undetermined (Gershwin 2006).

The Queensland species is very similar in appearance and habits to *Chironex*. Although smaller and less dangerous, it is still a severe stinger. The smooth bell is about 8 cm high and wide with a branched pedalium at each corner, each bearing up to nine tentacles which are shorter and thinner than those of *Chironex* and round in section (unlike the ribbon-like tentaces of *Chironex*).

Chiropsella sp. also differs from *Chironex* in the shape of the pedalial canal which is curved or bent at a right angle in contrast to the horn-like projection from the acute angled bend in *Chironex*. There are eight smooth rounded gastric 'saccules' and leaf shaped immature gonads with smooth edges quite unlike those in *Chironex* (Figure 6).

Chiropsella-like specimens have been collected in north-western Australia at Walsh Point in Admiralty Gulf and at Broome, where they have caused the hospitalisation of several people. *Chiropsella* has been found close to shore but not in creeks and rivers. A second species, *C. bart*, has been found at Gove in the Northern Territory during the dry season. It has a mild sting.

The venom has similar effects on the skin, blood and muscles to that of *Chironex*, but is less potent having about 1/100 to 1/200 of the total venom volume of a full-grown *Chironex*.

INJURY PREVENTION

- Wear lycra stinger suit, wetsuit or neck-to-toe clothes when diving, swimming or even launching boats in inshore northern waters (north and east of North West Cape in Western Australia) during the stinger season, September to May, and all year round in the far north.

SYMPTOMS

Stings from up to a metre of tentacle from small specimens are fairly minor and have little general effect. A more extensive sting causes severe pain and shock. General symptoms are similar to those of *Chironex* but less severe and are not life threatening.

- Several lines of weals surrounded by redness may develop. Weals have a pattern of bars corresponding to rings of stinging capsules on tentacles.
- Small blisters form along tentacle markings but there is only shallow tissue destruction and so no permanent scarring even though healing takes a long time.
- Skin discoloration, at first brownish and later purple usually disappears within six months.
- Area may itch at times as long as marks persist.

TREATMENT

- Wash with vinegar to deactivate nematocysts and prevent further stinging.
- Do not use alcohol (methylated spirits or concentrated drinks such as vodka, whisky, etc.) or fresh water.

Refer to treating mild to moderate coelenterate stings, page 23–24.

If general symptoms develop, seek urgent medical attention.

ROUND OR SAUCER JELLYFISH

Phylum Coelenterata
Class Scyphozoa
Order Semaeostomeae

The scalloped-edge round jellyfish (scyphozoans), commonly known as saucer jellyfish, have tentacles and sense organs around the edge of the bell or saucer and a single central mouth surrounded by four, large, mouth or oral arms that fall in long frilly folds. The life cycle includes a minute polyp stage attached to rock, that buds off a succession of tiny jellyfish, or ephyrae.

Patrick Baker

▲ The north-western Australian hairy stinger *Cyanea mjobergi*, photographed off Dampier, Western Australia.

FIGURE 7:

A simplified composite diagram showing the main differences between four genera of saucer jellyfish. In all but *Aurelia* the tentacles are shown as being shorter than in life.

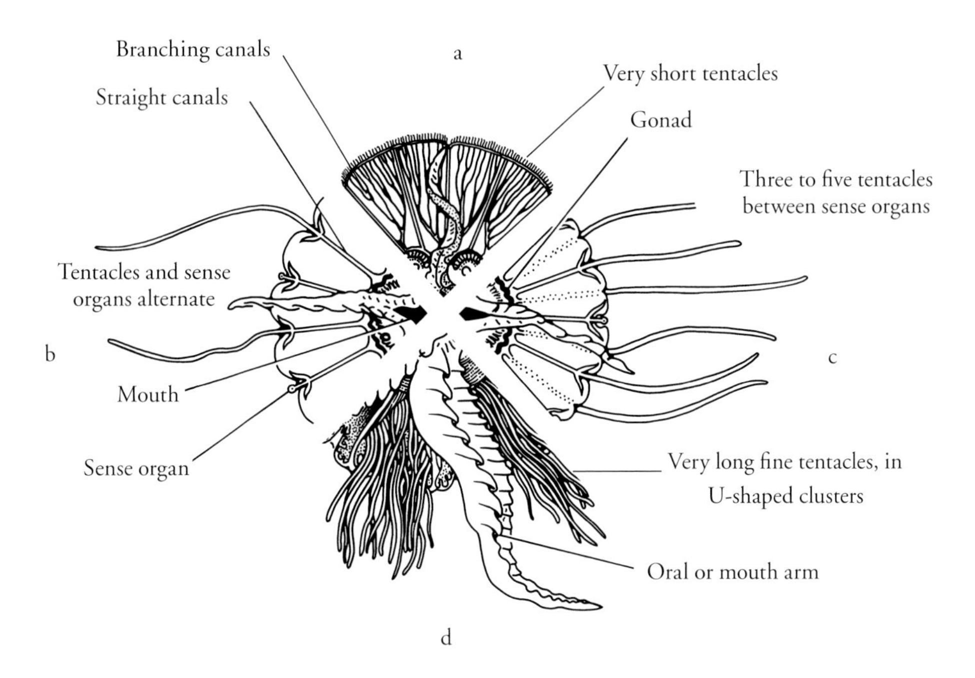

a) *Aurelia* has a fringe of very numerous short tentacles around the edge of the umbrella and branching canals leading from the stomach to a ring canal near the margin, short oral arms.

b) *Pelagia* has alternating tentacles and sense organs, with a total of eight tentacles, long oral arms.

   Sanderia has alternating tentacles and sense organs with a total of 16 tentacles.

c) *Chrysaora* spp. have three to five tentacles between sense organs with a total of 24–40 tentacles. *C. kynthia* has 24 tentacles, long oral arms.

   Pelagia, Sanderia and *Chrysaora* have tentacles arising singly from the margin, simple canals and no ring canal.

d) *Cyanea* has very long thin tentacles arising in U shaped clusters under the umbrella and long oral arms.

The best-known species of saucer jellyfish is *Aurelia aurita*, common worldwide in oceans, estuaries and inshore waters. This species is usually harmless although there have been reports of mild stings from England and Egypt.

Stinging saucer jellyfish found off Western Australia include the mauve stinger *Pelagia, Sanderia*, the sea nettle *Chrysaora* and the hairy stinger *Cyanea*. The sea nettle and the hairy stinger occasionally enter the Swan River estuary. The main differences between *Aurelia*, *Pelagia*, *Chrysaora* and *Cyanea* are illustrated in Figure 7.

TREATMENT FOR MILD TO MODERATE ROUND (SAUCER) JELLYFISH STINGS

Specific treatment, where known, is given for a species. Where no specific treatment is suggested, the recommendations below can be followed.

- Do not use vinegar, which could enhance stinging activity of nematocysts in round jellyfish.
- Never rub the skin, which could spread the nematocysts and, by applying pressure, cause nematocysts to discharge.
- For pain relief, wrap solid ice or frozen gel with layer of towel (or cooling could be too severe) and apply. Crushed ice in a plastic bag and cool cans of soft drink etc. can be safely applied to the stung area. Assess pain at 15 minute intervals and reapply cold pack if necessary.
- Applying a thin paste of bicarbonate of soda may ease pain.
- Aluminium sulphate as an ointment or spray may ease symptoms for stingings over a small area, and so may Aluminium chlorhydrate, found in deodorants.
- Applying the juice of a green pawpaw or the crushed leaves of the beach morning glory may help (see pages 19–20).
- Keep stung area rested, and if possible and relevant, elevated.
- For persisting pain, use a local anaesthetic such as lignocaine, as an aerosol, gel or ointment, or general analgesics such as paracetamol, aspirin and codeine.
- If there is no relief from pain or other symptoms develop, seek medical attention.

Mauve stinger

Pelagia noctiluca (Forsskål, 1775)

Phylum Coelenterata
Class Scyphozoa
Order Semaeostomeae
Family Pelagiidae

The bases of the tentacles of the round mauve stinger, *Pelagia noctiluca*, alternate with the sense organs around the umbrella margin. The stinger has eight sense organs and eight tentacles, and the shallow umbrella is usually about 4 to 6 cm wide and 3 cm deep (see below and page 69).

Four long mouth or oral arms hang from the centre; these and the umbrella are spotted with nematocyst warts. The mauve stinger varies in colour from brownish yellow to light rose-red, mauve or purple. The colour is strongest on the nematocyst warts, gonads and tentacles. As the specific name *noctiluca*, or night light, indicates, the stinger lights up when stimulated by touching or by wave turbulence.

The mauve stinger is an oceanic species, found worldwide in warm and temperate waters. It usually occurs in large swarms that are occasionally driven inshore during storms where they may cause mass stinging of

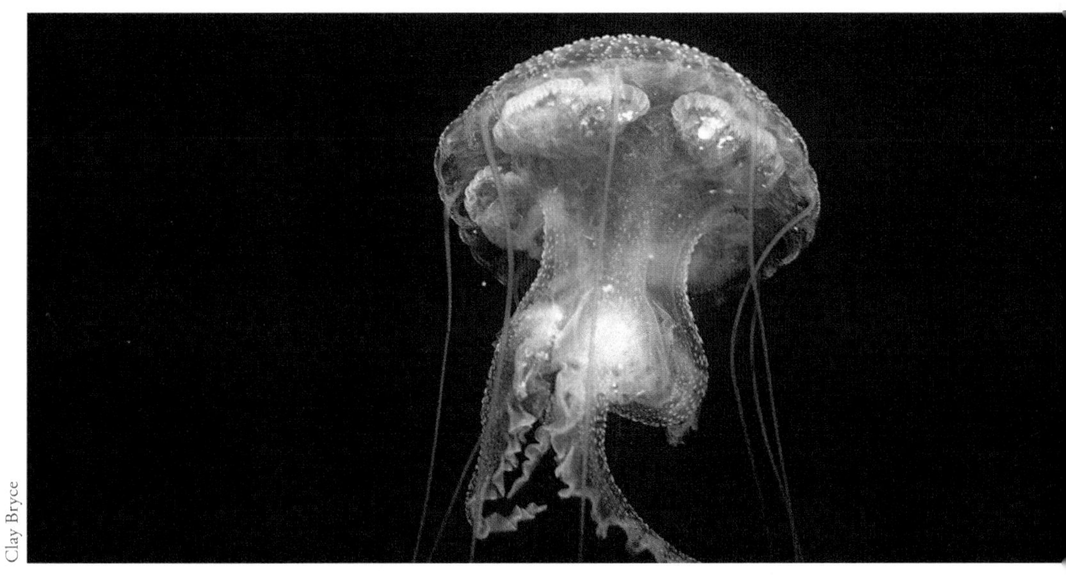

Clay Bryce

▲ The mauve stinger, *Pelagia noctiluca*, has a warty bell, long frilly mouth arms and only eight tentacles.

swimmers. Large numbers were reported in bays near Cape Naturaliste during 1984, and it has been found off Shark Bay and the north-western coast of Western Australia. The mauve stinger can be expected to occur anywhere along the west coast of Australia and from north Queensland to New South Wales on the east coast.

The venom of the mauve stinger is composed of a variety of proteins, and is antigenic (causes reaction of the immune system) in animal experiments, haemolytic (damages membrane of red blood corpuscles and allows haemoglobin to escape) and dermonecrotic (causes death of skin) in humans.

INJURY PREVENTION

- Avoid swimming if any jellyfish are washed up on the beach.
- Wear close-fitting cotton or lycra shirt with high neck and long sleeves when swimming.

SYMPTOMS

Contact with outer surface of umbrella or with trailing tentacles results in a mild sting, although multiple stings may be severely painful. A major allergic reaction (anaphylaxis) has been recorded in persons sensitised by other jellyfish stings, and so patients with a tendency to allergies should be monitored carefully.

- Immediate sharp pain is followed by intense itching that may persist for more than a week.
- In some cases there are small, raised, red lumps that may blister and fester, healing slowly.
- Extra pigmentation of skin in area can last some time.
- Multiple stinging is very occasionally followed by faintness, nausea, vomiting, difficulty in breathing and shock.
- Secondary infection sometimes occurs.

TREATMENT

- Maretić *et al.* (1987) advocated the use of a topical cortico-steroid anaesthetic preparation for *Pelagia* stings.

Refer to treating mild to moderate round jellyfish stings stings, page 70.

Refer to treating infections and allergies, page 219–223.

Seek medical attention if general symptoms develop after multiple stinging.

Sanderia sp.

Phylum Coelenterata
Class Scyphozoa
Order Semaeostomeae
Family Pelagiidae

A species of *Sanderia*, possibly undescribed, has been found off Port Hedland in Western Australia. The umbrella is round and flat, about 9 cm across. Its outer part falls vertically with 16 marginal tentacles, each alternating with a sense organ as in *Pelagia noctiluca*, but *Sanderia* has twice as many tentacles as *P. noctiluca*. Around the margin are 32 cleft marginal lobes (lappets). The upper surface of the umbrella has a number of large nematocyst warts. Under the bell are four, frilled, mouth arms which are longer than the diameter of the bell (Figure 7).

At present, no data is available on the nature of the venom of this species or on measures that can be taken to deactivate nematocysts. However, although it differs in some respects from the better-known *S. malayensis*, its stinging properties are likely to be similar. The tentacles of *S. malayensis* are glutinous and extremely sticky, suggesting a continuing discharge of nematocysts.

INJURY PREVENTION

- Wear wetsuit if diving offshore or lycra stinger suit if swimming in open water.
- Do not touch the bell of *Sanderia* sp.

SYMPTOMS

- *S. malayensis* sting is severe with local skin death.

TREATMENT

- Remove adhering tentacles by winding around stick or similar.

  Refer to treating mild to moderate round jellyfish stings stings, page 70.

  Seek medical attention if symptoms persist.

Sea nettle

Chrysaora spp

Phylum Coelenterata
Class Scyphozoa
Order Semaeostomeae
Family Pelagiidae

Species of the sea nettle *Chrysaora* are found worldwide but individual species may have quite restricted distributions. *C. kynthia* (Gershwin and Zeidler 2008), is so far known from off Perth and Fremantle metropolitan beaches and is particularly common in Cockburn Sound and Owen Anchorage during the summer and occasionally found in the Swan River estuary.

Chrysaora kynthia has a bell flatter than a hemisphere, up to 112 mm diameter, with colourless warts over the surface. There are 24 tentacles in eight groups of three between eight sense organs (Figure 7c, page 69). The tentacles are laterally flattened throughout their length, are very long and carry a formidable armament of nematocysts. Below the bell hang the

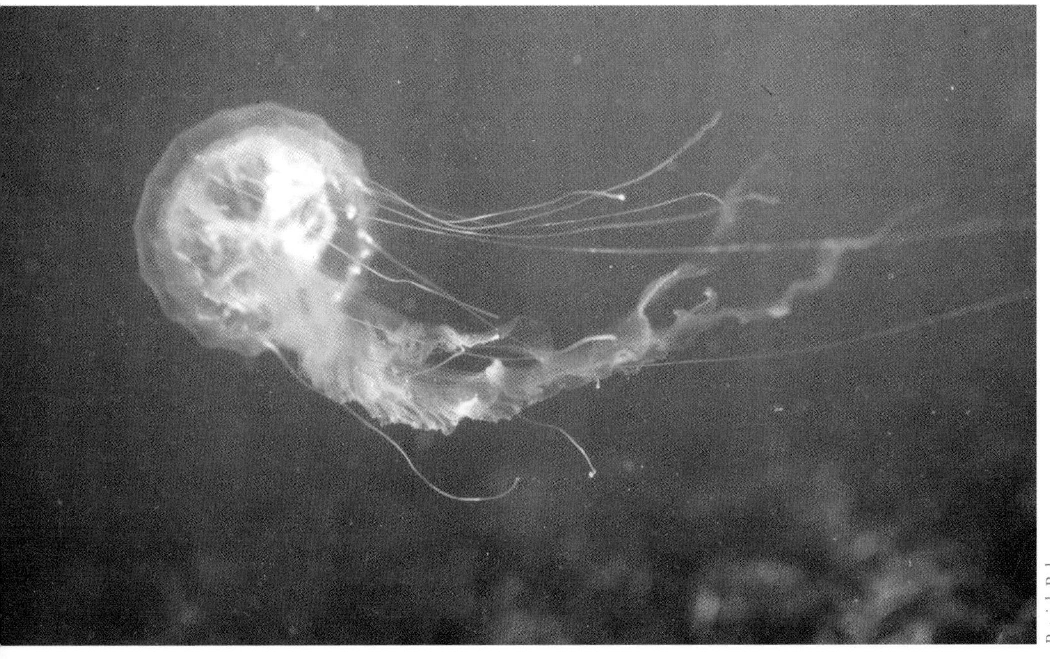

Patrick Baker

▲ The delicate bluish *Chrysaora kynthia*, often found in Cockburn Sound, is a severe stinger. The sting is relieved by application of wet baking soda.

long delicate frilly mouth arms about twice the bell diameter in length.

The colour in life is bluish-white without any pigmentation or pattern.

In South Australia another newly described species *Chrysaora southcotti* has red-pigmented nematocyst warts, densest in the centre of the upper surface of the cloudy whitish bell and scattered towards its margin. The oral arms are very dark reddish in colour. The bell is smaller than that of *C. kynthia* but carries 40 tentacles in eight groups. Both species are painful stingers.

INJURY PREVENTION

- Wear a wetsuit when diving, and a lycra stinger suit when swimming in open water.
- Avoid swimming if *Chrysaora* have been seen.

SYMPTOMS

The nematocysts of *Chrysaora* cannot penetrate deeply enough to deliver venom to the bloodstream so that the sting, which can be extremely painful, is not life-threatening.

- Sudden burning sting which persists for many hours.
- Sting is followed by raised red weals and spreading red area which takes two to three days to disappear.

TREATMENT

- Baking soda (sodium bicarbonate) powder applied to wet skin becomes a slurry which prevents further discharge of most nematocysts and relieves pain immediately. If pain persists, reapply bicarbonate of soda, which can also be mixed to a thin paste before applying.
- Papain ointment partially prevents further discharge of nematocysts but should not be rubbed in.
- Hot water treatment is not helpful.

Refer to treating mild to moderate round jellyfish stings, page 70.

Hairy stinger, or snotty

Cyanea spp

Phylum Coelenterata
Class Scyphozoa
Order Semaeostomeae
Family Cyaneidae

The name 'snotty' for the round or saucer jellyfish of the genus *Cyanea* comes from the mucus which is copiously secreted by the mouth arms and which has a fishy odour. The upper surface of the hairy stinger, or snotty, resembles that of the harmless common saucer jellyfish but there the resemblance ends. Numerous, long, thin tentacles (giving rise to the name 'hairy stinger') hang from eight U shaped areas on the underside with eight sense organs on the margin. Surrounding the central mouth are the greatly divided, frilly, mouth arms, which, with the tentacles, have been said to resemble 'a mop hiding under a dinner plate' (see figure 7).

There are a number of species of hairy stingers, *Cyanea capillata* is found worldwide in temperate and cold seas, and it seems that none of the *Cyanea* in Western Australia belong to this species. In other parts of Australia, *Cyanea capillata* grows to one metre wide whereas it can be three times as large in cold European and sub Antarctic waters, with tentacles 60 m long.

C. mjobergi is known only from north-western Australia, and may be the same species as *C. buitendijki* recorded from Darwin. It is about 14 cm across.

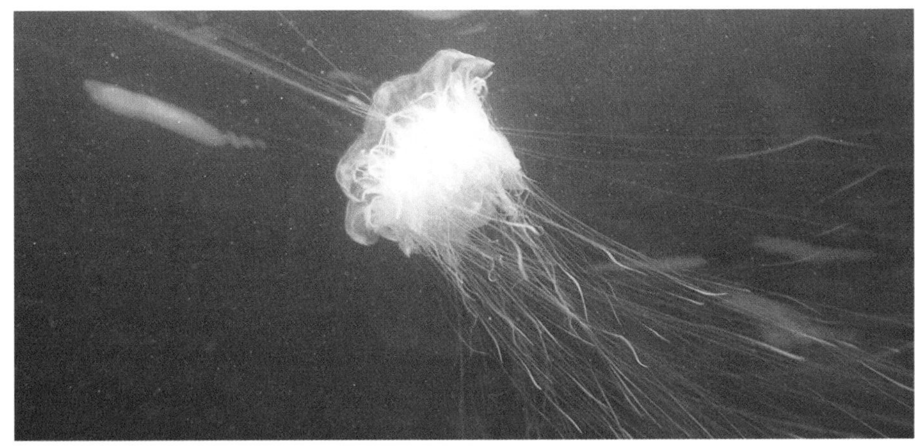

Clay Bryce

▲ A species of *Cyanea* photographed off Marmion Marine Park, Western Australia.

Most *Cyanea* in Western Australian waters are 15–20 cm across the umbrella with more than 1,000 tentacles, each capable of stretching to 3 m or more in length. Large specimens are occasionally washed up on beaches on the south coast of Western Australia. Pilots of light aircraft have sighted large *Cyanea* in Queensland waters.

Cyanea is uncommon in Western Australia but small hairy stingers, about 10 cm across, are sometimes found in the Swan River estuary; these have a slightly yellowish umbrella with yellow-brown mouth arms.

The stinging ability of *Cyanea* varies. Small *Cyanea* (10 cm across the umbrella) sting only mildly or may even fail to sting, when touched. Large *Cyanea* sting severely and a massive sting from a large specimen can be dangerous. Detached pieces of tentacle or tentacles trailing metres away from the body of the jellyfish are almost invisible and hard to avoid. Even dried tentacles on fishing nets are capable of stinging when remoistened. A venom composed of multiple proteins has been isolated from a species of *Cyanea*, and affects the heart of experimental animals. Some fishers are reported to have developed immunity against *Cyanea* stings although this is rare with marine venoms.

INJURY PREVENTION

- Wear a wetsuit when diving, and a lycra stinger suit when swimming in open water.
- Wear goggles for underwater swimming.
- Avoid swimming if *Cyanea* have been seen.
- Very fine, very long *Cyanea* tentacles, or pieces of them, can be encountered far from the bell.

SYMPTOMS

- A straight-line sting can occur from a piece of detached tentacle but more often, because of large number of fine tentacles, weals are multiple but narrow. They are often in parallel lines or zigzags, since tentacle may attach to victim's skin at point of initial contact and then contract across skin. A tangle of tentacles might cause a bizarre pattern of weals and a more severe sting.
- Weals are raised and white with a surrounding red flare.
- Pain varies with amount and length of time of contact with tentacles.

- Pain lasts an hour or more.
- Swelling fades rapidly but red marks can last for days; blistering is rare.
- If swimming underwater with eyes open and unprotected, contact with *Cyanea* tentacles can cause eye injuries. Pain is immediate and severe; eyelids swell, and superficial or deeper lesions to cornea can occur. With treatment these usually heal within a week without any residual effect. Nevertheless, eye injuries should be seen by a specialist.

General symptoms are uncommon but a severe sting could result in:

- Great difficulty in breathing.
- Victim may note that their heart is racing or apparently missing a beat.
- Severe chest pain could develop.
- Profuse sweating, mucus secretion from nose and throat, and muscle spasms have been described.

TREATMENT

- Stings to eyes should be treated by an eye specialist (see Case report).

Refer to treating mild to moderate round jellyfish stings, page 70.
Seek medical attention for general symptoms.

CASE REPORT

During the summer of 1960–61 *Cyanea* was responsible for many stings to swimmers at beaches of Port Phillip Bay, Victoria, Australia.

Among the victims of eye injuries were three young men who swam underwater with their eyes open, and remembered swimming into *Cynaea*. A fourth man was stung in the eye at night.

Pain was immediate and extremely severe with one patient requiring morphine for relief.

There was moderate swelling of the eyelid but the main injury was to the cornea with superficial finely punctate or linear epithelial abrasions or deeper lesions in which linear thread-like punctures penetrated to one third of the thickness of the cornea.

Antibiotic ointment and an eye pad were applied; in one case local homatropine and hydrocortisone were also given. All lesions healed within a week without any residual effect.

(Cleland & Southcott 1965)

RHIZOSTOME JELLYFISH

Phylum Coelenterata
Class Scyphozoa
Order Rhizostomeae

The bell of a round rhizostome jellyfish has a scalloped edge without marginal tentacles and a denser, more solid, umbrella jelly than does a saucer jellyfish. There are 8 mouth arms, greatly subdivided, with numerous tiny mouths instead of a single one at the centre. Tentacle-like filaments hang from the mouth arms in some species. None of the Australian rhizostomes are severe stingers.

The rhizostome jellyfish most common in Western Australia is the spotted jellyfish, *Phyllorhiza punctata* (see page 80). The net-patterned jellyfish, *Pseudorhiza haeckeli*, is not seen very often but is noticed because of its coloration. *Cassiopea*, the upside-down jellyfish, is found on sandy tropical reefs, and swarms of *Crambione* sp. and *Lobonema* sp. are sometimes seen in north-western Australia.

TREATING MILD TO MODERATE RHIZOSTOME STINGS

No specific chemical has been found to deactivate rhizostome nematocysts.

- Do not apply vinegar or alcohol (such as in methylated spirits or concentrated drinks such as vodka, whisky, etc.) or use fresh water as these are likely to cause further discharge of nematocysts.
- Never rub or wash the skin vigorously, which could spread the nematocysts and, by applying pressure, cause nematocysts to discharge.
- Remove obvious sticky pieces of animal.
- Apply ice pack or wrap solid ice or frozen gel with layer of towel (or cooling could be too severe) and apply. Crushed ice in a plastic bag and cool cans can be safely applied to the stung area. Assess pain after 15 minutes and reapply if necessary. Reassess after a further 15 minutes.
- For persisting pain, use a local anaesthetic such as lignocaine, as an aerosol, gel or ointment.
- If there is no relief from pain or if other symptoms develop, seek medical attention.

Spotted jellyfish

Phyllorhiza punctata von Lendenfeld, 1884

Phylum Coelenterata
Class Scyphozoa
Order Rhizostomeae
Family Mastigiidae

The round spotted jellyfish, *Phyllorhiza punctata*, is brown with white spots on its umbrella. The smooth yellow-brown umbrella of dense jelly may be up to 50 cm across although it is usually much smaller. It has

Clay Bryce

▲ The spotted round jellyfish, *Phyllorhiza punctata*, is a mild stinger, abundant in southern Australian estuaries and embayments.

8 sense organs but no tentacles around the edge, and the 8 extremely branched mouth arms, hanging from the centre, have tentacle like filaments hanging from them. The brown colour is due to single-celled algae that live in the jellyfish tissues and are passed down from the female parent in the egg. The jellyfish obtains much of its nourishment from these algae which also release oxygen.

The fertilised eggs of this jellyfish adhere the female's mouth arms where they grow into swimming larvae. The larvae then settle on the seabed and develop into polyps which bud off minute jellyfish or ephyrae.

The spotted jellyfish occurs in large numbers in sheltered bays and estuaries in south-western Australia, and is particularly abundant in the Swan River estuary. It is also known from eastern Australia, the Philippines and Japan. It was first named and described from Port Jackson, New South Wales, in 1884, but had been described, unnamed, from the Swan River estuary in 1837. The spotted jellyfish can often be handled without giving a sting but it is capable of mildly stinging sensitive parts of the body.

A related species, *Phyllorhiza pacifica*, is found in Shark Bay and off the north west coast of Western Australia. It has not been reported to sting.

INJURY PREVENTION

- Do not throw jellyfish at a person's face.
- No protective clothing is needed.

SYMPTOMS

- Immediate pain if stung in sensitive part of body such as eyes.

TREATMENT

Refer to treating mild to moderate rhizostome stings, page 79.
If eyes are stung, seek medical attention.

Upside down jellyfish

Cassiopea sp.

Phylum Coelenterata
Class Scyphozoa
Order Rhizostomeae
Family Cassiopeidae

Commonly known as the upside-down jellyfish, *Cassiopea* sp. is often seen on shallow sandy areas of coral reefs in north-western Australia. The bell is flattish and circular, about 8 cm across with a deeply scalloped margin and eight fringed much-divided mouth arms, and the colour may be brown, grey or greenish.

Its upside-down habit allows the absorbtion of sunlight for photosynthesis by the symbiotic algae in its tissues, supplying a large part of its nutrition. Small planktonic organisms taken in by the many tiny mouths on the frilly mouth arms supplement its food needs.

Cassiopea ndrosia has been recorded from Queensland and South Australia and *C. andromeda* from the Northern Territory. In Australia these are very mild stingers but the venom of *C. andromeda* from the

Clay Bryce

▲ *Cassiopea* sp. lies almost inert upside down on sand, absorbing sunlight for photosynthesis.

Red Sea causes the death of skin cells and damages red blood cells. The Western Australian populations of *Cassiopea* species have not yet been identified.

INJURY PREVENTION

- The upside down jellyfish is usually found lying still on shallow reef flats and so is easily avoided.
- Wear boots when reef walking.

SYMPTOMS

- Severity of stinging varies between species from no sting to moderate sting with weals.

TREATMENT

Refer to treating mild to moderate rhizostome stings, page 79.

Red rhizostome jellyfish

Crambione mastigophora Maas, 1903

Phylum Coelenterata
Class Scyphozoa
Order Rhizostomeae
Family Catostylidae

Every few years, swarms of the red rhizostome jellyfish, *Crambione mastigophora*, are found near the coast of north-western Australia. At times they have blocked the cooling water intake of power stations and are a considerable nuisance to swimmers.

The hemispherical bell is generally about 12 cm in diameter, edged with numerous scallops and has 8 rather solid, highly divided, mouth arms. The mouth arms have fine filaments about 15 cm long hanging from their extremities, and shorter ones near their upper ends. The bell is reddish brown with red gonads visible.

In March-April 2007, vast numbers of these jellyfish were seen within and outside the lagoon of Ningaloo Reef, off Cape Range Peninsula, Western Australia, and a few found their way south to Rottnest Island by June, carried there by the Leeuwin current. They had also been present near the Ningaloo Reef in April 1987 and 2000.

INJURY PREVENTION

- Avoid swimming if many of these jellyfish are seen.
- Wear a lycra stinger suit for protection, or a wetsuit if snorkelling or diving.

SYMPTOMS

- Immediate intense pain, subsiding after 4–5 hours.
- Redness with some swelling.
- No residual marks or scars.

TREATMENT

- Stings to eyes should be treated by an eye specialist.

Refer to treating mild to moderate rhizostome stings, pages 79.

◀ A swarm of *Crambione mastigophora* at the surface near Exmouth, Western Australia.

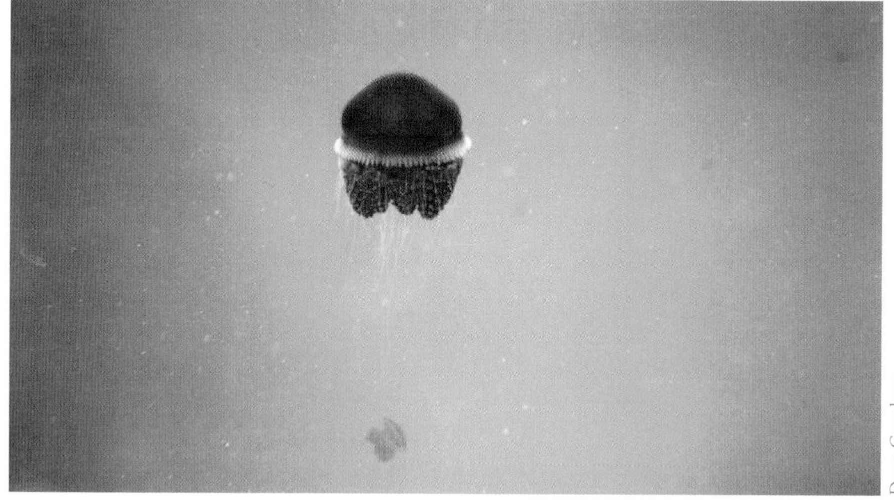

Doug Coughran

▲ *Crambione mastigophora*, a painfully stinging rhizostome jellyfish, off the Ningaloo Reef, Western Australia.

Giant rhizostome jellyfish

Versuriga anadyomene (Maas, 1903)

Phylum Coelenterata
Class Scyphozoa
Order Rhizostomeae
Family Versurigidae

The oceanic giant rhizostome jellyfish, *Versuriga anadyomene*, is rarely seen in Australian waters but is known from off the Great Barrier Reef and also off the Ningaloo Reef in north-western Australia. It is the largest known rhizostome jellyfish and can grow up to 60 cm across the umbrella. The umbrella is almost hemispherical, with a brownish network of surface grooves. In between the grooves are pale elevations.

Roger Steene

▲ The giant rhizostome jellyfish, *Versuriga anadyomene*, 25 cm long, photographed off Ningaloo Reef, Western Australia.

The margin of the umbrella bears 8, large, semicircular lobes alternating with small narrow lobes 1 pair in each of the 8 sectors of the margin. There are 8 marginal sense organs. The mouth arms are three-winged and broad, with flat membranous branches bearing small club-shaped bubbles, or vesicles, and some filaments. The mouth arms and filaments have a solid appearance due to the closely packed branches.

INJURY PREVENTION

- This is a very large conspicuous species and so easily avoided.
- Wear a lycra stinger suit or a wetsuit when diving offshore.

SYMPTOMS (WILLIAMSON ET AL. 1996).

- Painful sting lasting six hours.
- Ulcers may form at the sting site.
- Scar tissue and marks were evident after three weeks.
- No general effects have been recorded.

TREATMENT

- Do not use vinegar, alcohol or fresh water.
- Stings to eyes should be treated by an eye specialist.
- Refer to treating mild to moderate rhizostome stings, pages 79.
- If ulcers form, seek medical attention.

Net-patterned jellyfish

Pseudorhiza haeckeli (Haacke, 1884)

Phylum Coelenterata
Class Scyphozoa
Order Rhizostomeae
Family Lychnorhizidae

The net-patterned jellyfish, *Pseudorhiza haeckeli*, is not seen very often but its striking coloration attracts attention. The umbrella is covered with soft nodules which are outlined by a fine copper red network. *P. haeckeli* can grow to about 25 cm across though most specimens are 10–20 cm. The mouth arms are finely divided but are more delicate than those of

▲ The net-patterned jellyfish, *Pseudorhiza haeckeli*, is a mild stinger found more often offshore than in estuaries. The reddish net like pattern on the umbrella and the single long appendage distinguish this species.

Phyllorhiza, and a single, long, three sided appendage hangs from one of the eight mouth arms. This appendage is a brilliant blue, tipped with salmon pink. Filaments from the mouth arms are short.

Pseudorhiza is found in southern Australian waters from Western Australia to Victoria. Though usually found offshore, it does occasionally enter the Swan River estuary.

SYMPTOMS

- Under certain conditions, this net-patterned jellyfish may give a mild sting persisting for half an hour but it is usually regarded as a non-stinger.

TREATMENT

- Do not use vinegar, alcohol or fresh water.
- Stings to eyes should be treated by an eye specialist.
- Refer to treating mild to moderate rhizostome stings, pages 79.

Lobonema mayeri Light, 1914

Phylum Coelenterata
Class Scyphozoa
Order Rhizostomeae
Family Lobonematidae

Lobonema mayeri has a thick, tough, rigid bell up to 24 cm across with erect papillae (outgrowths) up to 4 cm long on the upper surface. The papillae are largest and grow most abundantly in the centre of the bell and are loaded with nematocysts. In each of the 8 sectors around the margin of the bell are four lobes which are up to 10 cm long, tapering, pointed and tentacle-like. The mouth arms are up to 15 cm long, each with the lower part three-winged. Numerous long spindle-shaped appendages ending in tentacle-like filaments hang from the mouth arms, and are responsible for the stings.

This species has been recorded in Malaysia, the Philippines and in Darwin harbour, Northern Territory. In April 2003, *Lobonema mayeri* was found in abundance off the Western Australian Kimberley coast.

Cleland and Southcott (1965) reported that in the Philippines the sting was regarded as 'very unpleasant' but 'not dangerous'.

INJURY PREVENTION

- Wear a lycra stinger suit or wetsuit when diving or snorkelling in northern waters.
- Avoid swimming if these jellyfish are seen in the area.

SYMPTOMS

- Sting is similar to nettle sting with white and then red weals but no blistering.
- Pain is severe to moderate, lasting for a few hours.
- No general symptoms have been recorded.
- No skin death or scarring results from sting.

TREATMENT

- No specific treatment is known.
- Refer to treating mild to moderate rhizostome stings, pages 79.

Sea anemones and Corals

Phylum Coelenterata
Class Anthozoa

These 'flowers of the sea' are often brightly coloured and attractive but they should be treated cautiously as they are related to jellyfish and, like them, have nematocysts (stinging capsules). The sticky feel of anemones is due to their nematocysts adhering to the skin, and a few species are severe stingers.

Sea anemones (Figure 8) have a tubular body with a single cavity that combines the function of digestive and circulatory systems and also carries waste and reproductive products which pass out of the mouth, the body's only opening. Internal partitions support muscles and the gonads, and increase the area for food absorption. The nematocysts are borne externally on the tentacles and internally on special threads attached to the partitions. Although anemones often appear fixed to the rocks, most have a muscular basal disc which allows them to glide about.

Anemones should never be eaten as many species are highly poisonous. No specific therapy is known for anemone poisoning, which could be fatal.

Clay Bryce

▲ The tube anemone, *Pachycerianthus* sp., with its column encased in a tough fibrous tube, lives in sand or mud.

FIGURE 8:

A sea anemone cut away to show the body cavity which combines the functions of digestive and circulatory systems. Through it pass waste and reproductive products to and out of the mouth, the only opening.

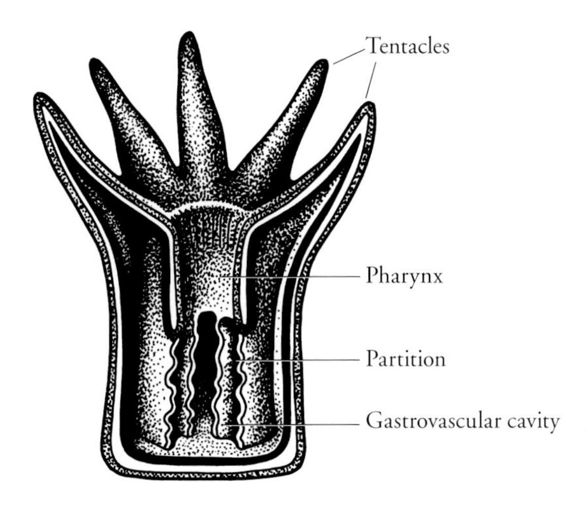

The tropical species *Heteractis magnifica*, *Entacmaea quadricolor* and *Stichodactyla mertensii*, all found on northern Australian coral reefs, contain toxins which can break down cell membranes and can be deadly if eaten. More than forty toxic peptides have been isolated from different species of sea anemones. Some sea anemones also produce neurotoxins in their tissues which lead to the blocking of nerve transmission leading to paralysis and respiratory and heart failure (Patocka & Strunecka 1999).

Another group of the class Anthozoa called zoanthids is related to anemones. Zoanthids do not sting but are among the most poisonous animals in the sea.

Stony corals are related to anemones but make a solid, fixed skeleton. They rarely sting but often are the cause of cuts that are slow to heal, and may be the cause of coral poisoning.

The fire corals, *Millepora* spp. (see page 35), common in northern Western Australian waters, belong to the class Hydrozoa and are only distantly related to true corals.

All anemones and their relatives should be treated with caution and gloves should be worn if handling them.

ANEMONES

Phylum Coelenterata
Class Anthozoa

Tube anemone

Pachycerianthus and *Cerianthus* spp.

Phylum Coelenterata
Class Anthozoa
Order Ceriantharia

The tube anemones, *Pachycerianthus* spp., are more closely related to the black corals than to true anemones. The long muscular but soft body is pointed at the bottom and enclosed in a tough fibrous tube up to 30 cm long by 3–4 cm wide, embedded in muddy sand or mud. When expanded, these animals have an outer crown of long, slender, mauve or white tentacles surrounding short thin tentacles around the mouth. In large groups they resemble a field of underwater flowers.

Some species occur between tides in mud flats in north western Australia but tube anemones are usually subtidal. A species of *Pachycerianthus* is common in Cockburn Sound and in the lower Swan River Estuary, at a depth of 5–25 metres.

A species of *Cerianthus* in New Caledonia is described as having stinging tentacles that 'inflict excruciating pain' (Catala 1964), but the stinging powers of related Western Australian species are not known.

INJURY PREVENTION

- Admire but avoid touching these anemones.

SYMPTOMS

- Intense severe pain.

TREATMENT

- Do not use vinegar, alcohol or fresh water as these are likely to cause further nematocyst discharge.
- Refer to treating mild to moderate coelenterate stings, pages 23–24.
- Seek medical attention if general symptoms, major allergic reaction or secondary infection develop.

Cockburn Sound anemone

Dofleina armata Wassilieff, 1908

Phylum Coelenterata
Class Anthozoa
Order Actiniaria
Family Actiniidae

The squat tubular body of the Cockburn Sound anemone, provisionally identified as *Dofleina armata*, is about 6 cm high by 4 cm wide, and lies partly buried in muddy sand or mud at a depth of between 1 and 20 m. The top of the body has a central slit like mouth and thick tapering tentacles around the edge, studded with oval beads heavily armed with potent nematocysts. The tentacles have a spread of 20 cm or more and wave actively, searching for prey as large as 10 cm long blowfish (*Tetraodon* sp.).

The body is light fawn with fine, dark brown vertical lines. The tentacles

Clay Bryce

▲ The large Cockburn Sound anemone, provisionally identified as *Dofleina armata*, has prominent beads containing nematocysts on the tentacles and oral disc and is a powerful stinger.

are either a rich to light brown above and below with white sides, or fawn with irregular bands of brown and white nematocyst beads.

Dofleina armata was described from Japan and is known from the Philippines, the Cockburn Sound anemone has been found from the tropical north of Western Australia to Shark Bay on the north-western coast, and in Cockburn Sound and the Swan River estuary near Fremantle in the south of the state. On tropical shores it has been found intertidally among mangroves.

INJURY PREVENTION

- Wear protective clothing when snorkelling or diving; the nematocysts of the Cockburn Sound anemone can penetrate light clothing.
- Be alert for Cockburn Sound anemones when swimming over muddy sand substrates and avoid them.
- Never wade among mangroves without boots that cover the ankles.

Neville Coleman

▲ *Dofleina armata* can spread to 10–20 cm across.

SYMPTOMS

- Intense burning pain.
- Weals patterned with white dead skin in the pattern of the batteries of nematocysts.
- Injury may be similar to full skin thickness burn, with correspondingly severe pain.
- In less severe sting, small blisters appear after five hours, with redness and swelling around stung area.
- A sting on the back of the hand produced scabs after 48 hours and then scars, still visible after two months.
- If skin is sensitive in stung area, ulcers may occur, taking several weeks to heal.
- Ulcers may develop secondary infections.
- General symptoms may be fever, nausea and vomiting, headache and, in very serious cases, shock.

TREATMENT

- Do not use vinegar, alcohol or fresh water as these are likely to cause further nematocyst discharge.
- Refer to treating mild to moderate coelenterate stings, pages 23–24.
- Seek medical attention if general symptoms, major allergic reaction or secondary infection develop.

Swimming anemone

Phlyctenactis tuberculosa (Quoy and Gaimard, 1833)

Phylum Coelenterata
Class Anthozoa
Order Actiniaria
Family Actiniidae

The swimming anemone, *Phlyctenactis tuberculosa*, is a large anemone, up to 15 cm across with a sac-like body up to 25 cm long. The body wall is covered by bubbles, or vesicles, which allow it to float, and it has numerous orange tentacles. The body colour is reddish-brown or bluish. The swimming anemone moves about by creeping on its basal disc or by actively swimming, and at night it moves high on plant fronds to catch floating prey.

This species is found in southern Australia from New South Wales to the south-west of Western Australia.

INJURY PREVENTION

- Wear protective clothing when snorkelling or diving.
- Avoid touching these anemones.

SYMPTOMS

- Contact with the tentacles will result in a moderately severe sting.

TREATMENT

- Do not use vinegar, alcohol or fresh water as these are likely to cause further nematocyst discharge.
- Refer to treating mild to moderate coelenterate stings, pages 23–24.
- Seek medical attention if general symptoms, major allergic reaction or secondary infection develop.

Barry Wilson

▲ The swimming anemone *Phlyctenactis tuberculosa* clings to algae during the day.

Loisette Marsh

▲ *Actinodendron* sp. spreads its tentacles flat on the sand and can retract quickly, leaving a hole about 4 cm wide. Photographed at Cockatoo Island, Kimberley, Western Australia.

Tropical stinging anemone

Actinodendron sp.

Phylum Coelenterata
Class Anthozoa
Order Actiniaria
Family Actinodendronidae

Tropical stinging anemones, *Actinodendron* spp., are large, grey-green or pale coloured anemones with 24 branching tentacles. They live subtidally in sandy areas adjacent to coral reefs and are able to withdraw completely into the sand leaving a hole about 4 cm wide.

Although a species of *Actinodendron* has been found in north western Australia southwards to the reefs of the Houtman Abrolhos, it is uncommon.

INJURY PREVENTION

- When snorkelling or diving, avoid touching this anemone.
- Wear lycra stinger suit or wetsuit.

SYMPTOMS

- In Queensland, *Actinodendron* is reported to cause very painful stings which remain sore and swollen for more than a week.
- Sting leads to a red rash of small blisters which in turn could lead to ulceration with delayed healing or skin death (Cleland & Southcott 1965), and a high incidence of abscess formation (Edmonds 1984).

GENERAL SYMPTOMS MAY DEVELOP (EDMONDS 1984):

- Fever, chills and thirst.
- Nausea and vomiting, with pain in abdomen.
- Headaches and delirium.
- Muscle spasms.
- Looking pale, and skin feeling cold and clammy i.e. shock.

TREATMENT

- Do not use vinegar, alcohol or fresh water as these are likely to cause further nematocyst discharge.
- Refer to treating mild to moderate coelenterate stings, pages 23–24.
- Seek medical attention if general symptoms, major allergic reaction or secondary infection develop.

Giant anemones

Heteractis and *Stichodactyla* spp.

Phylum Coelenterata
Class Anthozoa
Order Actiniaria
Family Stichodactylidae

The giant anemones, *Heteractis* and *Stichodactyla* species, have an undulating oral disc up to 80 cm across, covered in short tentacles. The tentacles of *Stichodactyla haddoni* are sticky to touch, and may fasten to human skin so strongly that they pull away from the anemone. *S. gigantea* has even more adhesive tentacles, but the nematocysts cannot penetrate human skin. At least two species of *Stichodactyla* are known to produce venom that causes rupturing of red blood corpuscles so any giant anemones should be treated cautiously.

There are several species of giant anemones on coral reefs in the Indo West Pacific and three, *Heteractis magnifica, H. aurora* and *Stichodactyla haddoni*, occur in north western Australia, the latter southwards to the Houtman Abrolhos (Fautin & Allen 1997). These species usually have associated anemone fishes.

INJURY PREVENTION

- Wear protective clothing and footwear.
- Avoid touching giant anemones.

SYMPTOMS

- Tentacles are strongly adhesive and can raise weals even if not painful.

TREATMENT

- Do not use alcohol (methylated spirits or concentrated drinks such as vodka, whisky, etc.) or fresh water.
- Refer to treating mild to moderate coelenterate stings, pages 23–24.
- Seek medical attention if general symptoms, major allergic reaction or secondary infection develop.

Clay Bryce

▲ The giant anemone, *Stichodactyla haddoni*, has an undulating disc about 30 cm across covered with very short, strongly-adhesive tentacles.

ZOANTHIDS

Phylum Coelenterata
Class Anthozoa
Order Zoanthidea

Zoanthids are anemone-like polyps a few centimetres high usually joined at the base or completely fused to one another, forming colonies. They do not sting but at least some species of *Palythoa* are highly poisonous.

Palythoa spp.

Phylum Coelenterata
Class Anthozoa
Order Zoanthidea
Family Zoanthidae

Sand grains embedded in the outer wall of the column give *Palythoa* a sandy colour and texture. The polyps of *Palythoa heideri* are about pencil thickness, a few centimetres long, and when expanded are topped with green or brown oral discs fringed with very short tentacles. In *Palythoa densa* the polyps are fused into an encrusting mat-like colony dotted with polyp openings.

This worldwide tropical genus has several species in north western Australia while two, *P. heideri* and *P. densa*, occur further south on limestone platforms, particularly those at Rottnest Island and near Margaret River, where they generally form a mat covering several square metres.

The toxicity of Western Australian species has not been investigated but a highly poisonous substance, palytoxin, has been isolated from species of *Palythoa* in the Pacific islands and north Queensland. It has the greatest known toxicity of any marine poison. It can cause severe internal bleeding, primarily affecting the kidneys but also the cardiovascular, gastrointestinal and respiratory systems. Death of experimental animals was due to kidney failure and general internal bleeding leading to heart failure and death (Moore & Scheuer 1971). Gleibs and Mebs (1999) found palytoxin in animals living close to colonies of *Palythoa*, such as sponges, soft corals, gorgonians (sea-fans), mussels and crustaceans. Fishes of the

Clay Bryce

◀ *Palythoa* spp. are highly toxic zoanthids. In Hawaii a species of *Palythoa* was traditionally used to poison spear tips.

genus *Chaetodon* and the crown-of-thorns starfish feed at least in part on *Palythoa* and accumulate high levels of palytoxin, which can be passed on to animals feeding on them. All these animals have a high level of tolerance to the toxin.

INJURY PREVENTION

- *Palythoa* should not be handled.
- Take great care that a cut surface of *Palythoa* does not touch any cuts or scratches.

SYMPTOMS

- *Palythoa* poison acts particularly on kidneys, heart and circulatory system when introduced through damaged skin.
- If poison comes in contact with victim's mouth, lesions result.

TREATMENT

- No treatment is known.
- Apply expired air resuscitation if breathing stops (see pages 227–228), and cardiopulmonary resuscitation massage if heart fails (see pages 229–231).
- Transport victim without delay to hospital for intensive care.

CORALLIMORPHARIANS

Phylum Coelenterata
Class Anthozoa
Order Corallimorpharia

Corallimorpharians are solitary or colonial animals that look like anemones but are structurally like hard corals without a skeleton. Some are moderate to severe stingers.

In southern Australia, the jewel anemone *Corynactis australis* is often red or orange with conspicuous white knobs tipping the tentacles. Contact with the tentacles causes severe pain with subsequent burning and itching. This species has been seen at Albany in Western Australia but a related tropical species which may be a severe stinger has been found at the Houtman Abrolhos, and also is known from the Great Barrier Reef, Queensland.

Discosoma spp.

Phylum Coelenterata
Class Anthozoa
Order Corallimorpharia
Family Discosomatidae

Discosoma rhodostoma forms dense aggregations (collections) on reef slopes and walls in New Guinea, the Philippines and the Western Pacific. Similar aggregations have been seen carpeting a sheltered reef slope in the Easter Group of the Houtman Abrolhos although that species has not been formally identified.

The expanded disc is about 10 cm across, and its surface is crowded with short tentacles, giving it a shaggy appearance. It is drab coloured, greenish or brownish, and when partly contracted, fine white vertical stripes are visible on the body wall.

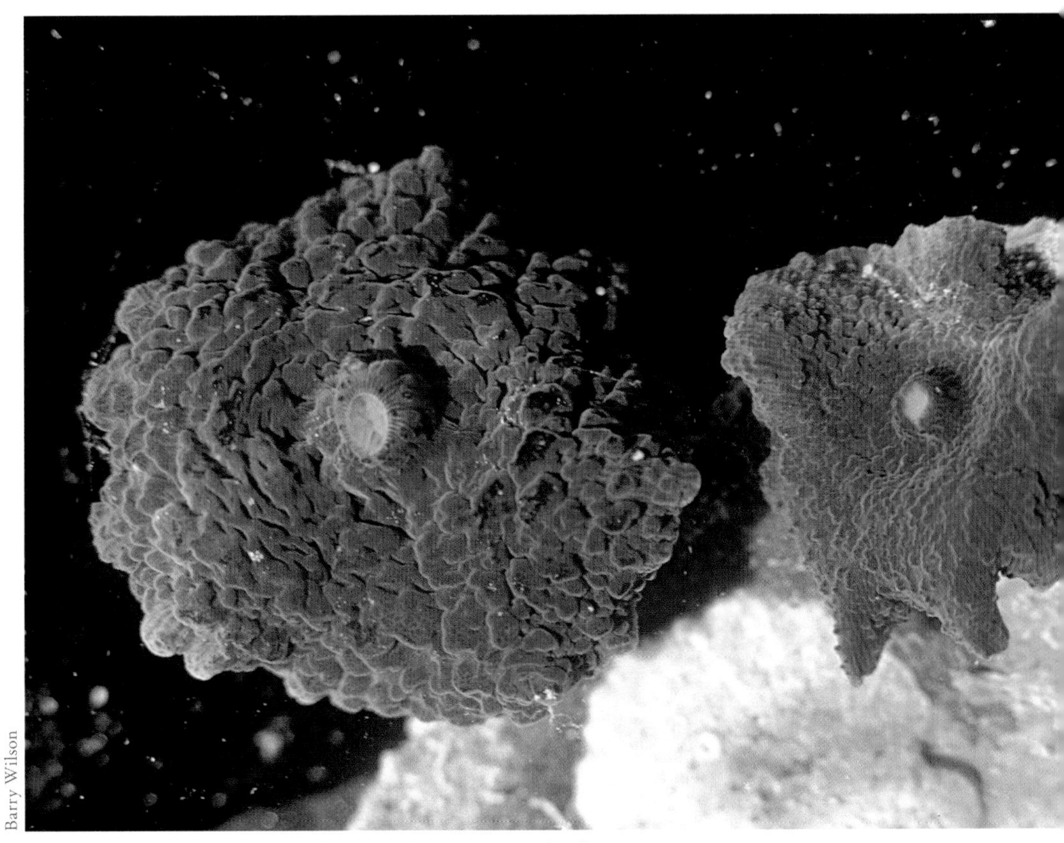

Barry Wilson

▲ Another species of *Discosoma* found at the Houtman Abrolhos.

INJURY PREVENTION

- When diving on coral reefs watch out for these rather inconspicuous animals carpeting the seabed, and avoid touching or kneeling on them.
- Wear a wetsuit — a lycra stinger suit does not give protection.

SYMPTOMS (FROM A CASE STUDY, WILLIAMSON ET AL. 1996)

- A sting on the knee through a lycra stinger suit by *Discosoma rhodostoma* was followed by raised red lumps over the elbow, knee and shoulder.
- Days later, incapacitating pain developed with pins and needles of the left arm and shoulder, and weakening of the arm.
- Complete recovery took eighteen weeks.

TREATMENT

- Do not use vinegar or alcohol (methylated spirits or concentrated drinks such as vodka, whisky, etc.).
- As the toxin of one species of corallimorpharian is deactivated by heat, try soaking stung area in water as hot as can be borne without further injury or scalding; use an unstung area to test temperature. Hot water soaking works fast and if successful, will avoid development of major symptoms.
- Do not use fresh water unless hot.
- Refer to treating mild to moderate coelenterate stings, pages 23–24.
- Seek medical attention if general symptoms develop.

STONY CORALS

Phylum Coelenterata
Class Anthozoa
Order Scleractinia

Stony corals have nematocysts for self defence and capturing food but most are not harmful to humans. Species of *Goniopora* and *Plerogyra* (Order Scleractinia) are among the few stony corals reported to sting. However, the greatest danger from corals is the chance of being cut by the sharp edges of their skeletons, especially if these are broken (see page 105). When people are walking on reefs, the coral often collapses underfoot, lacerating ankles and legs. Skindivers and surfers may be thrown against coral by waves but even brushing against corals can cause scratches that are hard to heal. This is due to the fragments of coral tissue including their symbiotic plants (zooxanthellae) and mucus from the coral that often remain in a cut. The body reacts against this foreign matter so subsequent coral cuts or scratches can cause a severe allergic reaction (see Allergies, page 221–223).

INJURY PREVENTION

- Wear boots when reef walking, and gloves when handling corals. Preferably, avoid touching corals.
- Be careful when stepping out of boats on to reefs and avoid stepping on plate corals, which are extremely fragile.
- Wear protective clothing and fins with heel protection when diving on coral reefs.
- If increasingly reactive to repetitive coral cuts, it may be necessary to keep away from coral.

Clay Bryce

▲ Plates of *Acropora* coral break easily and may lacerate the unwary diver or wader on a coral reef.

SYMPTOMS

It is most uncommon to encounter a pure sting from a true hard coral as the injury also involves cuts or scratches which require more attention than the sting.

- Small and often clean looking cuts may begin to sting a few hours later, especially during washing in fresh water.
- Mild inflammation may develop around the cut due to coral mucus, protein and possibly venom in the wound.
- Within the next day or two the inflammation may spread, with local swelling, redness and tenderness on light pressure and movement.
- Cut may develop into a festering sore.
- Lymph glands in the groin or armpit may swell and become tender and there may be pain or aching in adjacent joints.
- If properly treated symptoms will disappear after a few days but may continue for months if left untreated (see Fish handlers' disease page 220–221).
- Itching may persist for weeks or months.
- Allergic reactions may occur (see Allergies, page 221–223).

TREATMENT

- Keep an antiseptic with diving gear and apply as soon as possible after being scratched or cut.
- Scrub wound very gently with a soft brush and a bowl of fresh water containing a mild antiseptic or detergent.
- Do not use seawater. Scrubbing may cause discomfort and further bleeding but will hasten healing.
- Apply antiseptic.
- Cover with soft clean dressing and bandage.
- Injured part should be elevated and patient should rest.
- Oral aspirin or paracetamol may be used for pain relief.
- Seek medical attention for all but minor injuries, especially if the area is inflamed.

Clay Bryce

▲ The long polyps of *Goniopora*, expanded even during the day, completely hide the stony skeleton. Species of *Goniopora*, common on Western Australian coral reefs, can cause a mild sting releasing a toxin.

Clay Bryce

▲ *Plerogyra sinuosa*, found on coral reefs off north western Australia, is readily recognised by its inflated bubble like vesicles, expanded during the day. At night these are withdrawn and tentacles are expanded. *Plerogyra* can cause a moderate sting.

Flatworms

Flatworms, as the name implies, have a flat body. This is not segmented but has distinct head and tail ends, with a mouth but no anus. Within this large and diverse group of animals, those belonging to the class Turbellaria are free living, with some inhabiting damp forests while others live in freshwater and some in the sea. These larger marine flatworms are generally brightly coloured and secrete a defensive toxin from their skins.

However, most species of flatworms are parasitic and include the tapeworms (belonging to the class Cestoda) and the blood and liver flukes (class Trematoda). Several species of blood flukes, such as those belonging to the genus *Schistosoma*, may cause the serious tropical disease schistosomiasis in humans. These schistosome parasites infect and then multiply in species of aquatic snails, which are their secondary hosts. They then may infect humans or other species of mammals, which are their primary hosts. Fortunately this disease does not seem to have become established in Australia despite having been brought in occasionally by infected travellers returning from abroad. It is distressingly common in some African and Central and South American countries as well as in Madagascar and some south-eastern Asian countries.

Other genera of blood flukes that do occur in Australia normally infect birds after passing through marine or freshwater snail hosts. These schistosomes may accidentally infect humans, although they cannot then develop through their full life cycle. Instead, they die in the skin of the victim, causing a local irritation.

Clay Bryce

◀ Swan River sand flats at low tide (South Perth).

BLOOD FLUKE

Austrobilharzia terrigalensis Johnston, 1917

Phylum Platyhelminthes

Class Trematoda

Since the problem was first recorded in the early 1960s, bathers in the Swan River estuary have occasionally become infected by what is called swimmers' itch, or schistosome dermatitis. The cause of this condition has been traced to the blood fluke *Austrobilharzia terrigalensis,* whose adult stage normally lives in the common silver gull while its larval stages live in an estuarine snail, *Velacumantus australis* (below and Figure 9) (Appleton & Lethbridge 1979).

The eggs of this parasite are excreted in the bird's droppings and hatch into swimming larvae. These burrow into the soft body of the snail to grow there and multiply through several larval stages. The last larval stage, which looks rather like a minute tadpole and is known as a

C.C.Appleton

▲ The secondary host of the blood fluke, *Austrobilharzia terrigalensis*, is the estuarine snail *Velacumantus australis*. This snail is now common on sand flats in the Swan River estuary as well as estuaries along the eastern Australian coasts.

cercaria, emerges from the snail and swims to the water surface where it can survive for a day or two. If the larva then comes into contact with a silver gull it bores through its skin and migrates through blood vessels to the gull's liver. There it matures, mates and moves to veins around the bird's intestine. Its spiny eggs bore through the wall of the intestine and pass out with the bird's faeces to start the cycle again (Bearup 1956).

Schistosome dermatitis in humans occurs when these larval cercariae penetrate the skin, particularly the more tender skin of children. Because they are now in the wrong species of primary host, the cercariae cannot develop further and so die in the skin. However, for the human body, these cercariae constitute a foreign tissue and the body's reaction to it results in an inflammatory irritation (see page 113).

There are no really old records of schistosome dermatitis in the Swan because the snail *Velacumantus australis* is not a native to that area. It is believed to have arrived in the early 1950s from eastern Australia where swimmers' itch has long bothered swimmers in coastal lagoons. Such sheltered beaches, which are popular with families with young children, are unfortunately also the ones most favoured by both gulls and this species of snail (Pope 1972).

In the Swan River estuary, these snails are now abundant on shallow weed-covered sand flats between East Fremantle and Heirisson Island as well as in the Canning River as far upstream as about Mount Henry. The only other place in Western Australia where they have been known to occur is in similar habitats in the Woodman's Point area of Cockburn Sound.

Other types of swimmers' itch caused by cercariae of different species of blood flukes have resulted from exposure to the freshwater lakes of the wheatbelt of Western Australia and the lakes and rice fields of the Kimberley region. In these cases, the secondary hosts must again be species of freshwater snails and the primary hosts are probably aquatic birds other than seagulls, although the identities of these hosts are unknown.

The risk of swimmers' itch caused by *Austrobilharzia terrigalensis* in Western Australia is highest during the summer months, particularly when the water over shallow sand flats is at its warmest. Larvae are released from their snail hosts mainly during early morning and late afternoon.

Records indicate that most infection occurs in the late morning or among people prawning at night. As the snails, and so the larvae, occur close to shore in water generally less than a metre deep, there is less chance of infection when swimming into deeper water from a jetty than when wading in from the shore. It seems that larvae are not released from snails in the lower Swan estuary for a few weeks in mid-winter or in the middle estuary from about June to October when the salinity is lowered by the run-off from winter rain.

FIGURE 9:

The life cycle of the blood fluke, *Austrobilharzia terrigalensis*, which may cause swimmers' itch to bathers in the more estuarine reaches of the Swan River.

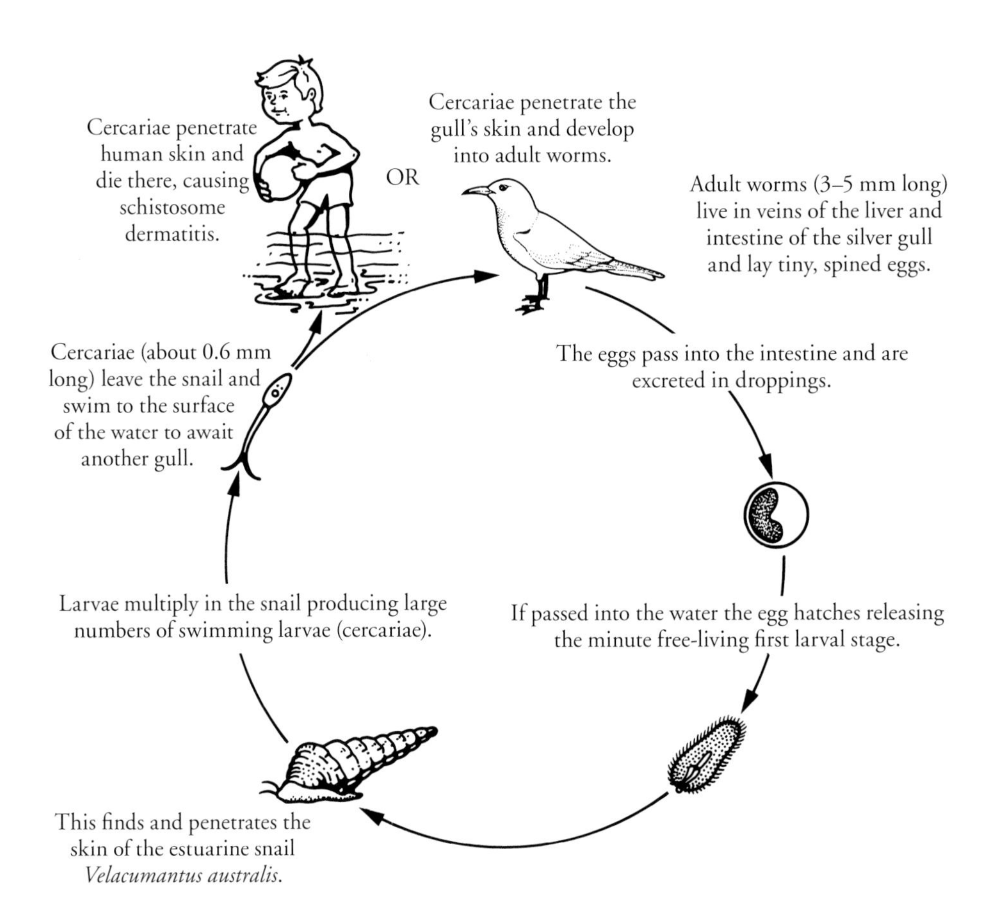

INJURY PREVENTION

- Avoid swimming or wading on river flats where the host snail is abundant and where seagulls concentrate.
- Rub the wet skin briskly with a towel immediately after leaving the water. This could remove some of the cercariae which have attached to but not yet penetrated the skin.
- If intending to spend much time in infected waters, wipe the skin with copper oleate (12.5% in yellow soft paraffin). Dimethylphthallate solution (25% in anhydrous lanolin) may also be used but this damages plastics such as vehicle upholstery.
- Wear long trousers tucked into socks or boots and a long-sleeved shirt when wading after crabs or prawns. This would also help to prevent injury from cobbler stings.

SYMPTOMS

- Half to one hour after contact a small red spot, about 1mm across, appears where each cercaria has entered the skin.
- Victim may have only a few spots or as many as 200.
- Spots become intensely itchy red lumps up to 5 mm in diameter, usually within 12–24 hours.
- Schistosome dermatitis only appears on parts of the body that have been immersed in infected water. Therefore, the distribution of spots in relation to clothing worn when wading or swimming, together with the story of water contact and then the long lasting nature of the patchy rash, usually allows a correct diagnosis. However, if the pattern is not obvious there can be confusion with chicken pox or impetigo (school sores).
- In severe cases, the lesions may look like eczema or may form blisters and there may be swelling, particularly of the ankles.
- Severe itching lasts for two to four days and the dermatitis for one to two weeks, or much longer if secondary infection occurs.
- Discolouration of the skin often persists for several weeks after the lumps disappear.

TREATMENT

- Calamine based creams, some of which contain a local anaesthetic, give some relief from the severe itching. These may be bought without prescription.
- Medical treatment may include anti-inflammatory and anti-allergic steroid ointments or antihistamine preparations and, if secondary infection develops, broad-spectrum antibiotics.

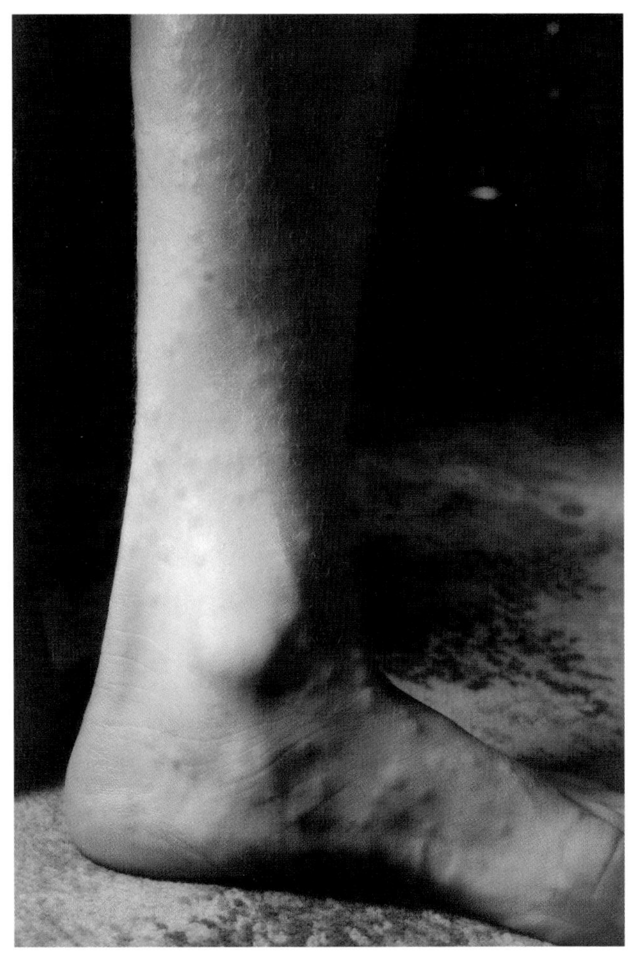

C.C.Appleton

▲ Swimmers' itch, or schistosome dermatitis, causes intensely itchy red lumps on those parts of the body that have been immersed in infected water.

SWIMMERS' ITCH FROM ALGAE

A different type of inflammation superficially similar to schistosome dermatitis can be caused by some blue green algae or cyanobacteria, such as *Microcoleus lyngbyaceus* and related species. These grow as a mat of fine brown or dark bluish green threads over the surface of sandy areas or may be attached to sea grasses or algae. They may be washed ashore by strong wave action.

Blue-green algal species are found worldwide but toxic effects have only been reported from the Ryukyu Islands near Japan and from Hawaii. They may have affected swimmers in Western Australia but, if so, this problem has not yet been positively recorded.

INJURY PREVENTION

- Avoid swimming where there are drift lines of detached seagrasses, algae and scum.
- Take a shower, preferably hot, as soon as possible after swimming, as the toxin from the algae is inactivated by heat.

SYMPTOMS

- Within several minutes to a few hours after swimming, people who have been exposed to blue-green algae and who remain in their bathers without taking a shower may suffer itching and burning sensations, followed by rashes, small blisters and finally, after four to six days, peeling skin.

TREATMENT

- Take a hot shower as soon as possible.
- Apply calamine lotion to the affected area.

Bristle worms

Phylum Annelida
Class Polychaeta

Worms of the phylum Annelida include earthworms, leeches and the marine bristle worms called polychaetes. All annelid worms have ringed or segmented bodies which are generally long and thin. Unlike flatworms they have 'straight-through' digestive systems with a mouth at one end and an anus at the other. They also have well-developed circulatory and nervous systems.

Most types of polychaetes live in the sea and are distinguished from other annelids in having a bristly projection, or parapodium, on either side of each body segment. These parapodia are muscular and well supplied with blood vessels. They serve as limbs and, in some species, also as gills. The bristles on the parapodia are fine and sharp and have a protective function as well as helping the animal to move along.

The polychaete fauna of our coasts is both abundant and diverse, and many of the larger sand-dwelling species are used as fish bait.

Most polychaetes are harmless but those that are particularly bristly, such as the spotted bristle worm, the fire-worm and the sea mouse, can cause problems. Some of the larger bait worms, such as *Eunice aphroditois*, can also give a nasty bite.

Clay Bryce

◀ A reef flat in the Kimberley will be home to many bristle worms, probably including *Eurythoe* and *Eunice*.

INJURY PREVENTION

- Be careful when turning over rocks and corals.
- Wear gloves when fossicking on reefs.
- Avoid touching any very bristly worms.

SYMPTOMS

- Contact with bristles initially produces a white, pinpoint rash and the area may become red, swollen and numb over 10–30 minutes.
- The site of the sting may become stiff, swollen and painful over the next 24 hours.
- Itching or burning may last for a week and numbness over the area may persist for several weeks.
- General symptoms may include increased pulse rate, palpitations, fainting and chest pain, causing the patient to feel very ill.

TREATMENT

- Bristles may be partly removed by applying adhesive tape to the area and peeling it off together with the bristles.
- Apply methylated spirits, calamine or other cooling lotions.
- Apply local anaesthetic ointment in more severe cases.
- Seek medical attention if the pain continues or general symptoms occur.

SPOTTED BRISTLE WORM

Chloeia flava (Pallas, 1766)

Phylum Annelida
Class Polychaeta
Family Amphinomidae

The beautifully spotted, oval, bristle worm, *Chloeia flava* (page 119), can grow up to 7 cm long. It lives in the silt on the seabed throughout the tropical Indo-West Pacific region, including the shallow waters along the northern Australian coasts. In Western Australia, it is found as far south as about Shark Bay, where it is moderately common under rocks and dead coral.

Clay Bryce

▲ The spotted bristle worm, *Chloeia flava*, has abundant fine bristles, which may be irritating to the human skin.

The spotted bristle worm is a carnivore and is often attracted to the bait on fishing lines, so is sometimes brought up when these are reeled in. Gloves should be worn when removing this worm from the hook as it has many fine bristles like those on some hairy cactus plants and these may become embedded in the fingers, causing varying degrees of irritation.

For Injury Prevention, Symptoms and Treatment, see page 118.

FIRE-WORM

Eurythoe complanata (Pallas, 1766)

Phylum Annelida
Class Polychaeta
Family Amphinomidae

The body of the fire-worm, *Eurythoe complanata* (page 120), is salmon-pink or grey with a thick fringe of white bristles projecting from the parapodia on either side of its body. A mature fire-worm may be up to 14 cm long and 1–2 cm wide including the bristles, but is commonly much smaller. Its fine chalky, or calcareous, bristles are hollow and if they

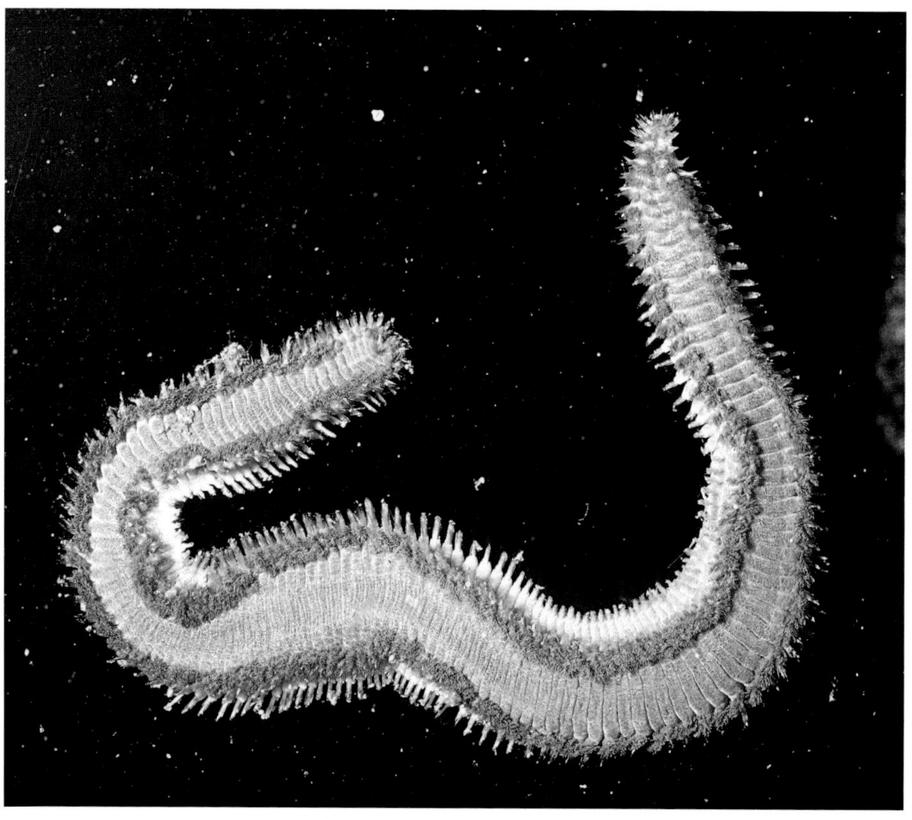

Barry Wilson

▲ The salmon pink fire-worm, *Eurythoe complanata*, has fine, hollow but chalky bristles which can cause blistering and numbness if they penetrate the skin.

penetrate the skin, blistering and numbness can result.

This species, living in the tropical waters of the Indian, Pacific and Atlantic Oceans, is abundant on coral or limestone reefs, where it shelters under boulders and in the weed mat. In Western Australia, it inhabits the shallow waters of the north west and west coasts and the offshore reefs and islands as far south as about Cape Naturaliste.

For General Injury Prevention and Symptoms, see page 118.

TREATMENT

- Flooding with vinegar will dissolve the calcareous bristles of *Eurythoe.*
- Apply methylated spirits, calamine or other cooling lotions.
- Apply local anaesthetic ointment.

Seek medical attention if the pain continues or general symptoms occur.

SEA MOUSE

Aphrodite australis Baird, 1865

Phylum Annelida
Class Polychaeta
Family Aphroditidae

The sea mouse, *Aphrodite australis*, may grow to about 15 cm in length. Its common name 'sea mouse' is due to its deep, broadly oval body, covered with fine brown bristles which are so dense that they resemble a coat of fur.

Dying and dead animals are sometimes washed up on beaches along the southern coasts of Australia and may be picked up because of their peculiar appearance. Stiff bristles projecting from and covering the parapodia and the sides and back of the body can cause problems when a sea mouse is handled without gloves. The penetration of these bristles may cause stinging similar to that inflicted by nettles, followed by an irritation that lasts for some hours.

For Injury Prevention, Symptoms and Treatment, see page 118.

Neville Coleman

▲ The sea mouse, *Aphrodite australis*, is found on silty sand in offshore waters of southern Australia.

Eunice aphroditois (Pallas, 1788)

Phylum Annelida
Class Polychaeta
Family Eunicidae

A very large species of bristle worm, *Eunice aphroditois*, can grow to be more than a metre in length and 2 cm wide. It lives in crevices or under boulders on tropical reefs, and can extend the front part of its body for some distance from its hole to forage on algae and debris that it bites off with black horny jaws. These jaws are large and strong enough to give an unpleasant bite to an unwary human (see page 122).

This species of polychaete worm is widespread in the tropical and subtropical areas of the Indo-West Pacific region, including the shallow waters of the north and west coasts of Western Australia.

INJURY PREVENTION

- Wear gloves and boots when fossicking on reefs.
- If handling *Eunice aphroditois*, seize worm behind its head.

SYMPTOMS

- Bite wound.

TREATMENT

- Wash wound area with antiseptic.
- Apply methylated spirits, calamine or other cooling lotions.
- Apply local anaesthetic ointment.
- Refer to treating infections and allergies, page 219–223.
- Seek medical attention if the pain continues or general symptoms occur.

Roger Steene

◀ *Eunice aphroditois* rearing from its shelter among rubble in search of food.

Crustaceans
Crabs, Prawns, Rock Lobsters, etc.

Phylum Arthropoda
Class Crustacea

Crustaceans, like insects, spiders and other arthropods, have jointed legs and a firm outer body covering which is sometimes called a shell. This covering serves as a skeleton — but outside rather than inside the body and so is more accurately named an exoskeleton. The upper part of the exoskeleton that covers the front part of the body of crustaceans such as crabs, prawns and rock lobsters is a single rigid structure called the carapace. However, more posterior sections covering the abdomen are jointed. The muscles, which are the meat we eat, are attached to the internal surfaces of these sections of the exoskeleton.

As crustaceans grow, they need to moult their 'old' exoskeleton. Before they do this they have to secrete a new one inside it. Initially this new exoskeleton is thin and soft but it becomes more rigid on exposure to seawater. The main parts of the exoskeletons of prawns and shrimps remain fairly flexible but crabs and rock lobsters further strengthen and harden their 'shells' internally with chalky, or calcareous, layers.

A crustacean uses its front legs and a series of smaller mouthparts to seize, tear and shred its food. Depending on its species, the food of a crustacean may consist of plants, animals or a mixture of both.

Apart from instances in which poisoning has been traced to spoilage due to bad cooking or to poor handling of prawns, rock lobsters etc, there

Clay Bryce

◀ Western rock lobsters, *Panulirus cygnus*.

are a few species of nut, or xanthid, crabs that have often proved to be poisonous even after correct handling and cooking. These species seem to accumulate toxins from their food. The crabs themselves are not affected by these toxins but their flesh then becomes poisonous to their predators, including humans.

Some people learn by sad experience that they cannot eat or even touch crayfish or prawns, or both, without developing an allergic reaction within a few hours (see Allergies, pages 221–223).

Clay Bryce

▲ Some people develop allergies to crustaceans such as this aggressive blue manna (or blue swimmer) crab, *Portunus pelagicus*, photographed in Cockburn Sound, Western Australia.

Clay Bryce

▲ Swimming crabs such as these blue mannas have clawed front legs for food capture and defense, and paddle-like back legs for swimming.

Clay Bryce

▲ Western rock lobsters, *Panulirus cygnus*.

Clay Bryce

▲ Western king prawn, *Penaeus latisulcatus*, in the Swan River, Western Australia.

NUT CRABS

Atergatis floridus (Linnaeus, 1767)
Zosimus aeneus (Linnaeus, 1758)

Phylum Arthropoda
Class Crustacea
Family Xanthidae

One of the distinguishing features of many species of the robust nut, or xanthid, crabs is the black tip of each 'finger' of their claws. Xanthids differ from some other types of crabs such as the blue manna sand crab, *Portunus pelagicus*, and the mud crab, *Scylla serrata*, in that they do not have any of their five pairs of legs flattened to form swimming paddles, but are well adapted as reef dwellers.

Shawl crab, devil crab and rough crab

Of all the species of xanthids, only three are positively known to be toxic. Two of these, the shawl crab, *Atergatis floridus* (page 129), and the devil crab, *Zosimus aeneus* (page 129), are common on the shallow tropical reefs of the coastal and offshore waters of northern Australia, including those of Western Australia. The third known poisonous species, the rough crab *Platypodia granulosa*, has not yet been found in Western Australian waters.

Of these three, the most dangerously poisonous species seems to be the devil crab, *Zosimus aeneus*, since it is the largest as well as having the greatest concentration of toxin in its flesh. Half a gram of the claw muscle of this species may contain sufficient poison to kill an adult human.

The toxin is thought to be saxitoxin, an extremely powerful type of paralytic shellfish poison (PSP), one milligram of which can kill about 5,000 mice. This toxin is initially formed inside the cells of certain dinoflagellates (single-celled planktonic organisms). If these dinoflagellates are filtered from the water and eaten by molluscs such as mussels and clams (see page 177–179), the toxin may accumulate in their flesh. In turn, these molluscs may be eaten by species of xanthid crabs, some of which are able to further concentrate the toxin. Apparently neither the molluscs nor the crabs are sensitive to the toxin but it may have a drastic effect on humans who eat these seafoods or even drink the broth in which they are boiled.

Clay Bryce

▲ This poisonous *xanthid, Atergatis* floridus, is commonly called the shawl crab because of the lacy colour pattern on its wide carapace.

Clay Bryce

▲ This devil crab, *Zosimus aeneus*, was photographed in shallow water at Scott Reef north of Broome, Western Australia.

PSP is not destroyed during human digestion nor by heat, which is unusual for marine toxins. However it is water soluble and so would be diluted during boiling (Llewellyn 1998). It is a potent neurotoxin that prevents the nerves of the victim, and so the muscles they supply, from functioning normally. Asphyxiation and death may occur if the muscles responsible for breathing are among those paralysed. Poisoning by xanthid crabs, as in any instance of poisoning by PSP, affects children much more seriously than adults.

Research work is being carried out in many laboratories throughout the world to develop methods of detecting and measuring the PSP content of animal tissues such as those of these xanthid crabs. A rapid method of assaying the saxitoxin content of dinoflagellates has been developed, but currently there are no antisera that have been developed for PSP poisoning.

POISONING PREVENTION

- Never eat any of the species known to be toxic.
- Avoid eating any crabs with black tipped claws.

SYMPTOMS

The severity of the symptoms depends on the crab's species and the amount of its flesh eaten. It also may vary between crabs of the same species, depending on the concentrations of PSP which have been passed up through the food chain, linking dinoflagellates, bivalves, crabs and humans.

- Within 15 minutes to a few hours after ingestion lips and tongue may tingle, and then burn or become numb. This feeling may spread to hands.
- Limbs become weak.
- A strange disorder of sensation may develop so that hot feels cold, and sweet tastes sour.
- Eye problems may develop — eyes may not be able to move, followed by double vision; eyes may not be able to focus, producing blurred vision; pupils may become fixed and dilated; and eyelids may not be able to blink so that they fail to protect the surfaces of the eyes, which may dry out. Unless eyes are shaded, these eye problems may bring the

added danger of too much light entering eyes, and focusing on and burning the retina.

- Dizziness, clumsiness, loss of coordination, difficulty with speech and swallowing, drowsiness, muscle paralysis, burning in throat and stomach, and even respiratory failure and coma could follow.
- In addition, severe vomiting and diarrhoea may follow, sometimes after salivation and thirst.
- Death may occur within 4 to 6 hours, even up to 12 hours, after ingestion and is due to respiratory paralysis.

TREATMENT

Direct treatment must concentrate on keeping respiration going. Seek expert medical attention as quickly as possible, but do not leave the victim alone.

- If breathing starts to fade, apply expired air resuscitation (EAR) (see pages 227–228).
- Maintain EAR until the victim is in a medical centre that can take over respiratory ventilation. Life support may be needed for up to 24 hours.
- Before paralysis, vomiting may be induced but do not induce vomiting after paralysis.
- Protect victim from choking on vomit.
- Protect victim's eyes.
- Other diners, even if showing no symptoms, may be induced to vomit as a preventative measure.

SEA LICE

Phylum Arthropoda
Class Crustacea
Order Isopoda
Family Cirolanidae

Some small isopods, which measure up to about 1 cm in length, belong to the group of crustaceans that, in general, feed on dead animals and plants and are known as sea lice. They closely resemble their counterparts on land, commonly called slaters and pill bugs, as their oval bodies are obviously segmented and flattened from top to bottom.

Clay Bryce

▲ These specimens of *Cirolana harfordi* are some of many individuals collected from the Swan River by Fremantle Port Authority divers after they had been attacked by a swarm of these isopods. The animals had entered the divers' wet-suits and inflicted painful bites.

Although the northern hemisphere species *Cirolana harfordi* is not native to the waters of southern Western Australia, it has been introduced and is reported as occasionally being abundant in that area. It is a scavenger on dead animal matter but is also capable of injuring humans and other animals. Divers working on a jetty in the Fremantle area reported that they had been 'attacked' and bitten by large numbers of these isopods which they had disturbed. One of the divers had an area of the skin of his chest deeply scraped by one of these animals that had crawled in under his wet suit (Pritchard 1991).

Another species of *Cirolana* in South Australia and a species of the related genus *Excirolana* in more tropical Australian waters have caused similar injuries to the legs of unwary waders in sheltered waters such as among mangroves. A similar 'attack' has occurred in the Sydney area from a species of *Actaecia* but, on that occasion, the victims were sunbaking on a dry sandy beach.

TREATMENT

- Clean wound and apply antiseptics locally.
- Use oral analgesics such as paracetamol for pain relief.
- Refer to Infections and Allergies, page 219–223.
- Seek medical attention if secondary infection develops.

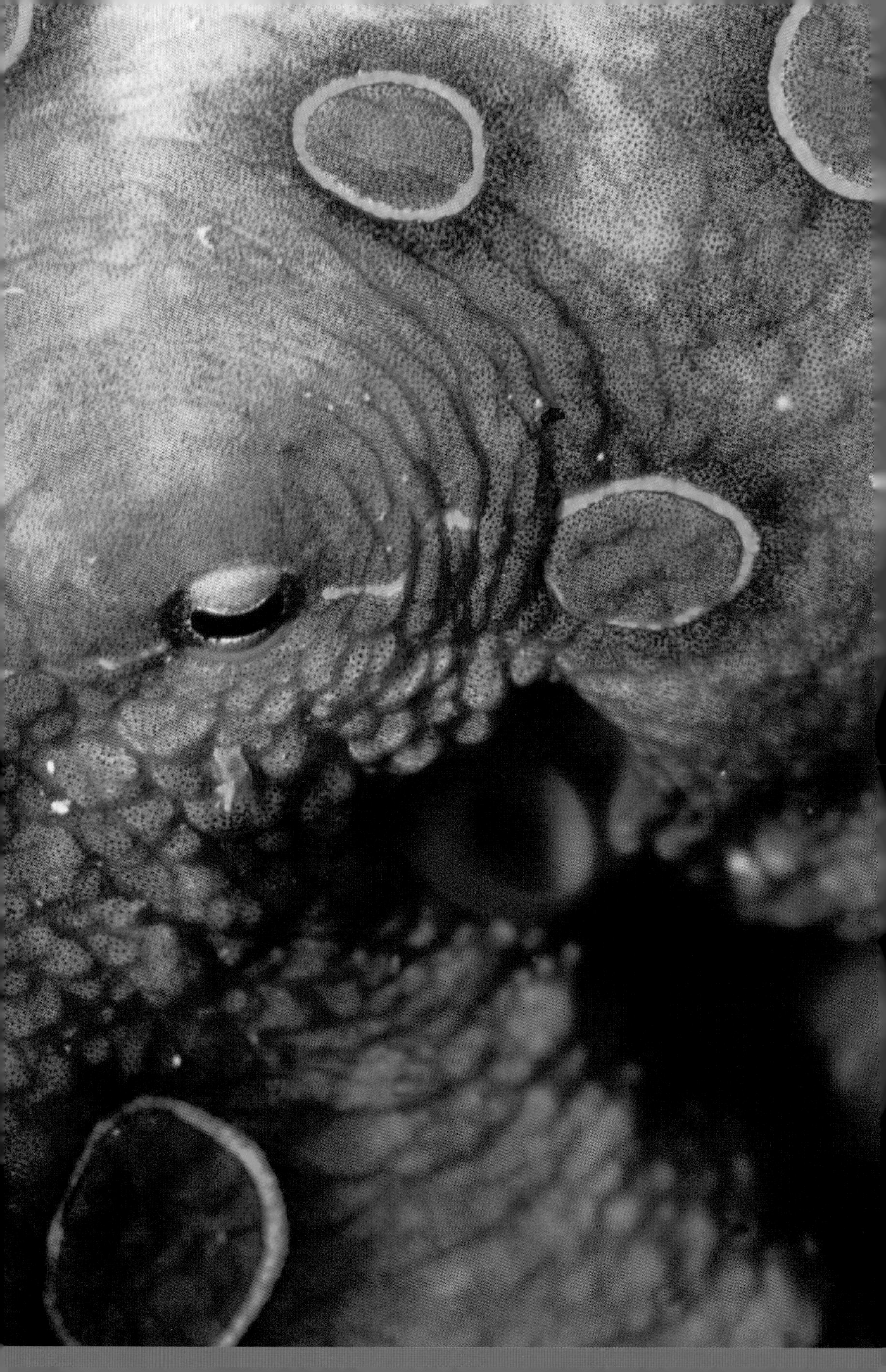

Molluscs

Cones, Octopus, Mussels, etc.

Phylum Mollusca

Dead cones which are washed up on the beach, oysters, squids, scallops and even garden pests such as snails and slugs are all molluscs despite the obvious variety of their body-forms and life styles. They range from fixed, heavy shelled, giant clams to streamlined, thin shelled, swimming scallops; from sea-dwelling abalone with their pair of large gills to air-breathing rams horn snails in freshwater ponds; and from slow moving, weed-eating limpets to swift squid which may even leap out of the water when chasing fish.

The phylum Mollusca includes a very large number of species, which have a huge diversity of body form. However, they all have soft bodies that are not divided into segments as are those of worms and arthropods. Typically, a mollusc's body consists of a head and a muscular foot. Attached to these is its visceral mass, containing the greater part of its digestive, excretory and reproductive systems. In most forms these systems open into and drain out through a space called the mantle cavity that generally contains the gills. This cavity lies between the visceral mass and a fold of skin called the mantle. The mantle can be muscular, as in squids, or membranous, as in air-breathing forms like most land snails in which the mantle itself may act as a lung. The soft bodies of most mollusc groups are protected by a hard shell, which is secreted by the mantle.

Although most molluscs have shells to protect their soft bodies, squid

Clay Bryce

◀ Close-up detail of eye and rings of *Hapalochlaena lunulata*.

and cuttlefish have only an internal vestige of a shell for muscle support, and octopus and some land and marine slugs do not have a shell at all. Among the most noticeable external features of most molluscs are the sense organs that are generally located on or near the head. Among these are the eyes and tentacles with which the mollusc can see, feel and taste its environment.

Muscular organs like the crawling foot of a snail move the mollusc towards food or a mate, or away from danger. In a bivalve such as a mussel, the largest muscles close the shell if conditions become unfavourable, while others pull on the 'beard', anchoring the mussel more closely to its jetty pile or rock substrate.

Some of the gastropod or coiled molluscs such as periwinkles, limpets and garden snails are herbivorous. They eat leafy aquatic or terrestrial plants, scraping them up with a ribbon of minute teeth called a radula and biting off pieces with a pair of horny jaws. Some other gastropods are

Clay Bryce

▲ Trying to right itself, this false ear shell, *Granata imbricata*, extends its muscular foot and exudes a milky fluid which is thought to deter predators.

carnivores, many with teeth and jaws specially shaped for catching and devouring their prey.

Bivalve molluscs, with shells comprised of two fairly similar valves, have neither jaws nor a radula. Most groups of bivalves use their large gills to sieve tiny particles from the water, a few eat minute living organisms from the plankton while others eat scraps of decaying plant and animal tissues.

There are two ways in which certain marine molluscs can harm and even kill humans. Some may cause illness and even death when eaten because of the poisons accumulated in their flesh. Others may cause injury by using the mechanisms with which they normally catch and devour their food. They use their radular teeth and, in some instances, their jaws to wound prey. The wound itself is usually not dangerous, particularly to larger animals such as humans. However, if toxic fluids are injected through the wound, more serious consequences may follow.

Cones

Phylum Mollusca
Class Gastropoda
Family Conidae

Cones are so named because of the more or less conical shape of their shells. Varying with the species, the sides of a cone shell may be straight or curved and convex. Its surface may be smooth, granular or ridged and is generally ornamented with a colour pattern. This shell pattern may be hidden by the horny outer layer of the shell, or periostracum, which may be thick and even hairy. The shell itself may be thick and heavy, or thin and lightweight to better suit a fast-moving animal.

Each species of cone has a characteristic size, shape, shell texture and colour pattern, which usually vary a little between individuals in that species.

Cones are both diverse and abundant in the tropical areas of the Pacific and Indian Oceans, particularly in the shallower waters of the continental shelves and on oceanic reefs. Fewer species inhabit the cold to temperate waters of the areas towards the poles and the deeper waters of the continental slopes and below. In the early 1990s more than 130 species of cones were recorded from Australian waters (Kohn 1998; Wilson 1994).

By 2009 the number of cone species recognised in the Australian fauna was 166 and 117 of these are known in Australian waters. (J. Singleton, personal comment, December 2009).

After mating the female cones attach many horny egg capsules to hard substrates, particularly the undersides of boulders. Within each capsule are numerous eggs, the number and size of these varying with the species. The period of development through to hatching lasts for between about one to three weeks, the degree of development of the hatchlings varying with the species and being related to the yolk content of the eggs. Some may hatch as planktonic larvae, others as crawling juveniles.

One of the most obvious features of an active cone's body is its siphon. This is a flexible split tube that can be extended and contracted, protruding from the edge of the mantle through a notch in the narrow anterior end of the shell just above the cone's head (see page 139). It is like a snorkel, the cone drawing a current of water through its siphon into its mantle cavity. There, this current of water passes over an organ which detects the scent of animals, including those on which the cone feeds. The water current then flows through the gill where carbon dioxide is exchanged for oxygen and out over the animal's right 'shoulder'.

Most species of cones prey only upon worms or molluscs, including other cones. However, some kill and eat small seabed-dwelling fishes. Like most predators of mobile animals, a cone has a very efficient sensory and nervous system, and can move rapidly. Sampling the water in different directions with its flexible siphon, it is able to home in on its prey by sensing the chemicals spreading from it through the water, even when the prey is moving.

THE HARPOON

Those cones that catch slow moving worms or molluscs seem to approach their prey fairly slowly. When near its prey, the cone extends its proboscis, a tube formed from the muscular lips of the mouth. The proboscis, usually contracted and kept tucked inside the snout, moves towards an appropriate area of the prey's body and then suddenly strikes, stabbing a tooth, gripped by the end of the proboscis, into the flesh of its prey. Some cones have been reported to shoot out a tooth like the dart from a blowpipe or the harpoon fired from a whale-chaser's gun.

Clay Bryce

▲ The colour pattern of the shells of some live cones may be obscured beneath a horny outer layer. This horny layer may even be covered with bunches of hairs, as in this geographer cone, *Conus geographus*.

Clay Bryce

▲ The snout and stalked eyes of this rearing Geographer cone, *Conus geographus*, can be seen below the long tubular siphon and above the more brightly coloured muscular foot.

The tooth of a cone is a slender hollow horny tube, generally with a barbed point. It is formed and then stored in a small sac that opens into the mouth cavity (Figure 10). The teeth of cones, unlike those of most other gastropods, are not joined in rows onto a basal strip (Figure 10) to form a rasping radula but are quite separate, though they lie in bundles inside the storage sac. When needed, a tooth is passed from this tooth sac through the mouth cavity and along to the tip of the proboscis. As the cone's tooth is plunged into the body of its prey, venom is injected into the wound. This venom has come from the venom gland through the mouth cavity, up to the tip of the proboscis and into the cavity of the hollow harpoon. The victim is paralysed rapidly by the venom from one or more teeth, and the muscular proboscis then widens out to encircle and engulf its immobile prey.

Because fishes are much more mobile than most worms and molluscs, the species of fish-eating cones need to act quickly to catch their prey. Even when buried under the sand, such a cone, with its tubular siphon

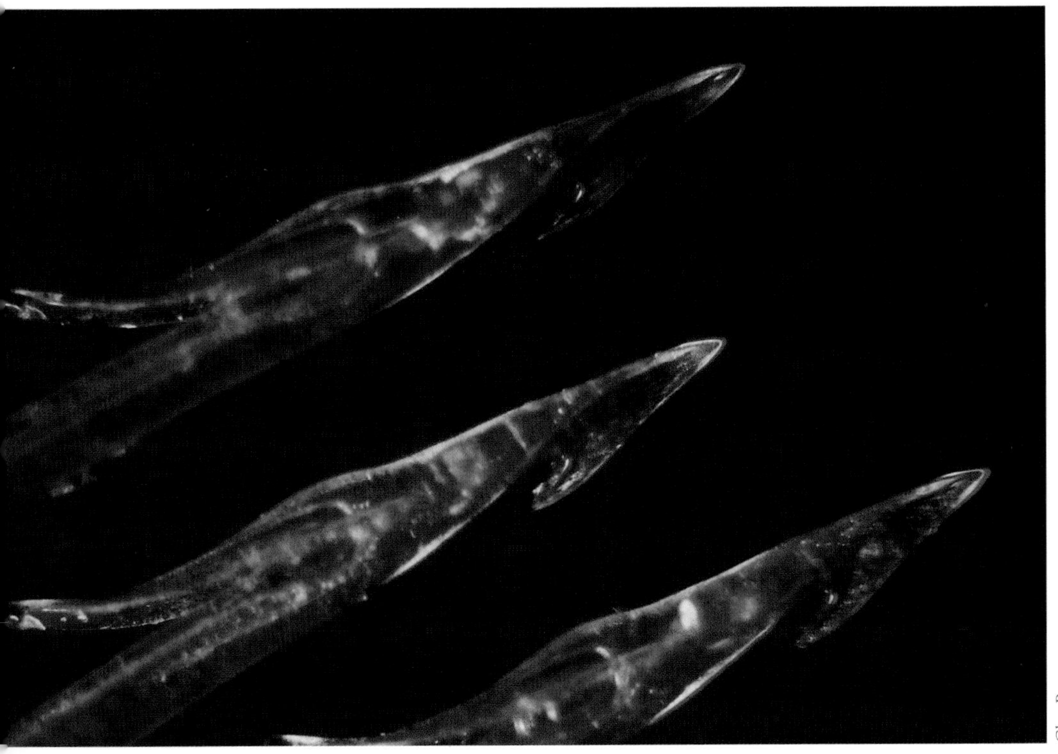

Clay Bryce

▲ The tips of three of the many teeth stored in the tooth sac of a striated cone, *Conus striatus*.

FIGURE 10:

Diagram of those organs in a cone's body that are associated with prey capture and feeding.

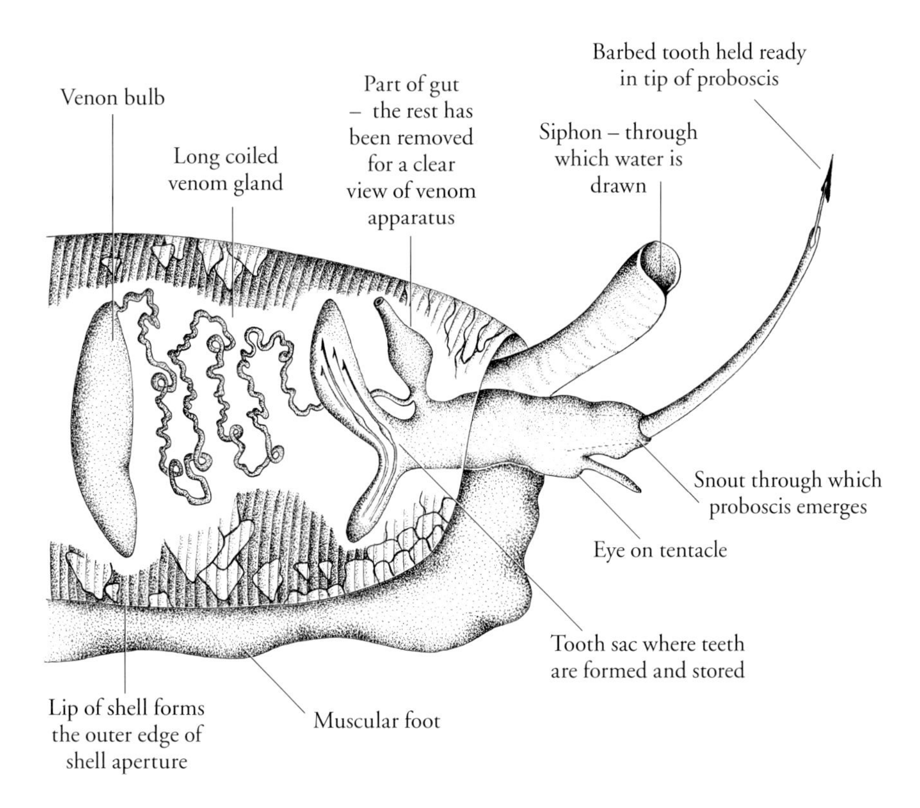

protruding from the sand like a snorkel, can track a fish as it swims nearby. It may leap out of the sand to plunge a tooth into the fish's body. On other occasions fish-eating cones are known to creep up on their prey with a slow gliding movement. In captivity, as when kept in an aquarium, some cones have been seen to engulf pieces of fish flesh without using their venom apparatus at all. They simply expand the proboscis and mouth and move up to envelop the food.

Fish-eating cones have been known to swallow prey of more than half their own size. The cone with the front part of its body greatly distended around a half-swallowed fish, usually retreats to shelter under the sand or into a crevice where it gradually digests its meal.

CONE STINGS

The behaviour of only a few of the 300 or so known species of cones has been thoroughly investigated. Most species seem to be active and to attack their prey only at night. However, their feeding activity may be governed more by hunger than by the time of day.

The potency of cone venom seems to vary not only from one species to another but also between individuals within a species. The potency of the venom of a single animal may also vary from one time to another. Those cone venoms that have been investigated consist of a mixture of more than one type of toxin. A temporal variation in potency might be due to these toxins maturing at different rates.

The species of cones that appear to be most dangerous to man are those which normally feed upon fishes. Most fish-eating species, such as the geographer cone, have long slender teeth with sharp cutting blades, which may ensure a deeper penetration of the venom. The venom injected through these long harpoons causes a variety of symptoms, including

Clay Bryce

▲ *Conus victoriae.* Great care is needed when handling a live cone as its almost transparent proboscis, armed with a tooth, can quickly lengthen and move in any direction.

numbness, swelling, inflammation and sometimes pain around the wound. General numbness and paralysis of the skeletal muscles, which are those attached to bones, can follow rapidly. Although the venom of some species may affect heart muscles and the central nervous system, the most general and important effect is on the skeletal muscles of the chest wall and of the diaphragm. Paralysis of these muscles can cause breathing to become difficult and even to stop.

The venoms of cones that eat worms or other molluscs contain a complexity of toxins including some protein -digesting enzymes. These enzymes may cause pain, cell disintegration, bruising, swelling and bleeding when injected into larger animals including humans. The venom of such a cone species does not seem to cause paralysis, even of the worms on which they prey, although it might slow them down. In general, the teeth of such cones are relatively short and robust with large knobs at their bases around which the tip of the proboscis apparently contracts to hold the tooth firmly.

PREVENTION AND TREATMENT OF CONE STINGS

All living cones should be regarded as potentially dangerous, and so should be handled as little as possible. Do not take risks with any cones, particularly with *Conus geographus* which is known to be extremely dangerous.

Collectors are advised to wear heavy gloves or to use a knife, stick or bent wire when dislodging cones from under rocks or within crevices. Live cones should be rolled over a couple of times so that the animals retreat into their shells. They can then be picked up but should always be held only briefly and at the widest part of the shell near the spire — the generally conical section of the shell behind the aperture. However, cones are very agile and can twist their bodies through almost 180° as they extend out of the shell. The proboscis can elongate right back to the shell's spire, so cones should be quickly placed in a collecting bag or pail. They should never be held in the hand, put in a pocket or placed inside a wetsuit.

Many cone injuries have been inflicted when a live cone is being carried in a pocket or fold of clothing or in a bag of thin material. The cone's attack reaction can be triggered off by pressure, rubbing or close confinement, particularly when a cone is crushed among other animals. The cone may then act like an animal backed into a corner, striking out at anything within reach.

Although only a few species of cones are known to inflict serious injury or cause death to humans, nobody knows how many other species might be dangerous. There also seems to be a great variation in people's susceptibility to the venom.

The different species of venomous cones cause a variety of symptoms in their human victims. There may be an obvious sharp pain where the victim has been 'stung', and the area around the puncture site may become inflamed, may seem to burn, and may swell and become numb. The more serious symptoms are weakness, clumsiness, difficulty in swallowing and speaking, the development of vision problems, and local paralysis, which may become general.

The most important symptoms are those associated with an increasing difficulty, and perhaps ultimately failure, of the breathing process. The onset of the symptoms can occur quickly, progress over several hours, reach a peak and then go away.

Treatment must be directed at keeping the victim alive through the period when the venom is having its greatest effect.

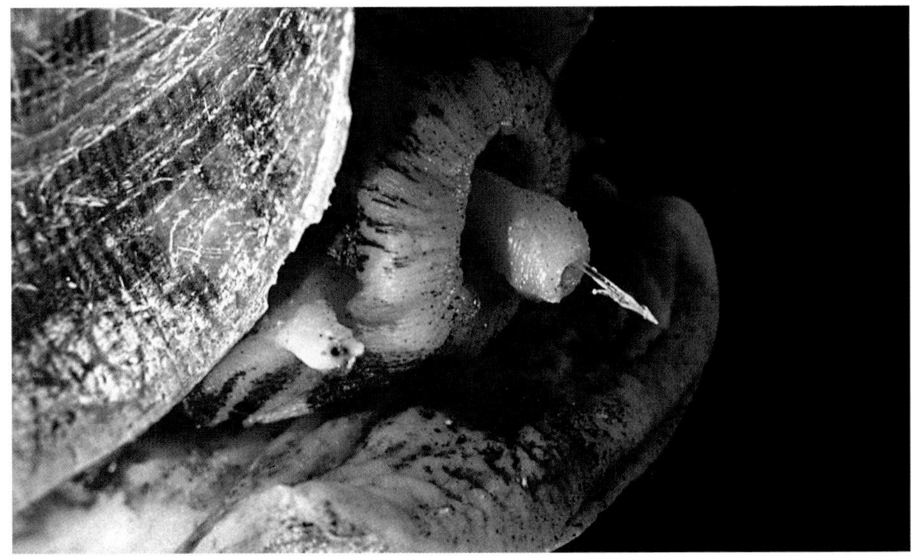

Clay Bryce

▲ This preserved specimen of a striated cone, *Conus striatus*, died as it was almost ready to strike. A tooth is held in its mouth at the end of its contracted proboscis.

FIGURE 11:

The distribution along the Western Australian coast of species of cones known or thought to be dangerous to humans.

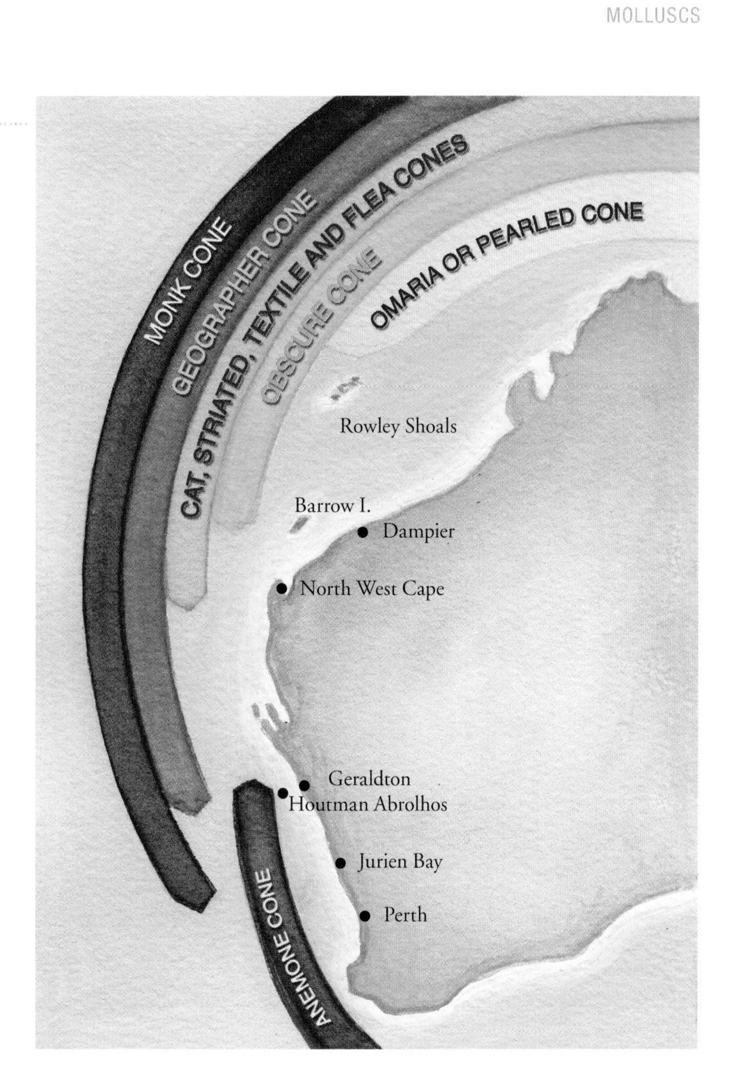

PAINKILLER EXTRACTED FROM CONE VENOM

University of Melbourne researchers have licensed a biotechnology company to commercialise a potential painkiller extracted from a toxic marine snail. The compound is known as ACV1 and was extracted from *Conus victoriae*, a beautiful but often deadly type of mollusc found in tropical waters near Broome, Western Australia.

Laboratory studies have suggested the drug could be more powerful and longer lasting than morphine, yet unlike morphine, non-addictive and lacking the side effects, namely sedation, nausea, constipation and respiratory depression.

UNINEWS, UNIVERSITY OF MELBOURNE, VOL.12 NO. 20, 17 NOVEMBER 2003, P. 7

INJURY PREVENTION OF CONE STINGS

- Wear heavy gloves when searching for cones and/or use an implement to dislodge them from crevices and from under ledges.
- Roll live cones over a couple of times so that the animals retreat into their shells before picking them up.
- Hold the live cone at the widest part of its shell well away from front end of the aperture and near the spire. Place quickly into a collecting bag or pail.
- Never hold a live cone for more than a few seconds and do not put it into a pocket or inside a wetsuit — a cone's tooth can penetrate layers of thick clothing.

SYMPTOMS OF CONE STINGS — LOCALISED INJURY, MODERATELY TOXIC VENOM

- Possibly an obvious sharp pain where the victim has been struck; the area around it may seem to burn and may swell and become numb.
- Pain can range from mild to extremely severe.
- Symptoms can occur quickly, progress over several hours, reach a peak and then go away.

SYMPTOMS OF CONE STINGS — POTENTIALLY DANGEROUS INJURY, HIGHLY TOXIC VENOM

- First, numbness and tingling which spreads from puncture to all over body, especially to the mouth, at high speed — in about 10 minutes. Pain may or may not occur.
- Weakness spreads rapidly from the puncture site throughout the body, making speech and swallowing difficult, and causing doubling and blurring vision.
- Further eye problems may develop: eyes may not able to move, followed by blurred vision; pupils may become fixed and dilated; and eyelids not able to blink, so failing to protect the eyes which might become dry.
- Increasing shallowness of breathing is the most serious symptom, and could progress to respiratory failure.
- Symptoms may become evident within 10 to 30 minutes, and progress over one to six hours.

TREATMENT OF CONE STINGS

Generally the danger develops in one to 6 hours, after which victim would probably start to recover.

- Bind the area around the wound with a pressure bandage as for snakebite; do not make too tight.
- If wound is to arm or hand, immobilise in a sling to delay spread of venom through rest of body.
- If the torso or leg has been punctured, apply a pressure pad as well as a pressure bandage and immobilise whole body of victim.
- Do not remove pressure bandage until victim has been brought to medical attention. When bandage is taken off there will be a very real risk that the sudden release of venom will have a drastic effect on the nervous system.
- Carry victim if it is necessary to move, especially if puncture is in body or legs.
- Put victim in the recovery position (lying on one side) to minimise danger of choking if vomiting should occur (see recovery position on pages 224–226).
- Never leave victim alone. Paralysis can cause breathing failure and vomited material may be inhaled and cause choking.
- If breathing becomes shallow and victim begins to turn blue, expired air resuscitation (EAR) is of paramount importance (see page 227–228). Maintain EAR until effects of venom wear off.
- If heartbeat fades, cardiopulmonary resuscitation (CPR) must be added to expired air resuscitation (see page 229–231).
- If eyes cannot move, provide shade to protect retinas from sun.
- Remember that the victim, even though paralysed, is still conscious and able to hear and perhaps to see. It is vitally important to realise that a victim can understand and will be affected by what is said, even if unable to communicate. Symptoms and treatment should be carefully explained to reduce stress and prevent panic; any comments on a victim's condition should be optimistic.

Darren Mok

▲ Geographer cone, *Conus geographus.*

Geographer cone

Conus geographus Linnaeus, 1758

The geographer cone, is a fish-eating species known to be responsible for at least twelve human fatalities as well as many serious injuries. This cone should never be touched and great care should be taken if it is encountered accidentally. The shell of this species is thin, smooth and lightweight. It has a low spire and is shaped more like a swollen cylinder than a cone.

Unless masked by the thick horny outer coat of the periostracum, the shell is basically pale blue or pale pink with a fairly strong pattern of red brown lines forming a network to enclose small triangular patches. This is called a 'tent pattern'. Running around the shell are irregular bands of brownish blotches formed by a denser or in-filled tent pattern. Inside the shell is white, although some of the exterior colour pattern may show through the thin shell, which can be up to 13 cm long.

This is the largest of the fish-eating cones and its venom contains a number of toxins, some acting directly on human skeletal muscles and at least one affecting the central nervous system. If paralysis occurs the muscles become limp. Severe localised pain and swelling may also occur.

The geographer cone is widely distributed through the tropical regions of the Indian and Pacific Oceans, and is found in Western Australian waters as far south as the reefs of the Houtman Abrolhos off Geraldton. It is not often seen alive because it hides in the sand under coral slabs and rock ledges during the day.

Dangerous injury, highly toxic venom

For prevention, symptoms and treatment of injury see pages 146–147

CASE REPORT

An experienced shell-collecting diver, having found and placed a geographer cone in a bag attached to his weight belt, was painfully stung through the bag, wetsuit and bathers. Fortunately he was near the boat and, with the help of his wife, was able to climb aboard. After being rushed to hospital he was treated over some hours until the symptoms began to abate. He is one of the few victims, if not the only one, to recover from a verified geographer cone sting. Possibly the tightness of his wetsuit restricted the spread of the venom, just as a pressure pad and bandage would have done.

MILTON AND AILEEN EAST (VERBAL ACCOUNT)

Cat cone

Conus catus Hwass in Bruguière, 1792

The shell of the cat cone, is rather small and stout, and is quite heavy for its size. The sides of the last-formed or body whorl are slightly curved, particularly near the raised, straight-sided spire at the posterior end of the shell. The texture of the body whorl is roughened by grooves and fine nodular ridges, which run around the shell. The colour of the shell is brown with irregular broken patches of white or pale blue, which tend to form radial bands along the length of the body whorl. These radial bands tend to intersect two bands of darker patches running around the body whorl. Near the anterior end of the shell there are white spots which, in some specimens, form narrow radial bands across the ridges. The shell can be up to 5 cm long.

The cat cone is said to feed on fish. It seems to have inflicted only minor injuries to humans but, despite its small size, should be treated with caution.

Clay Bryce

▲ Cat cone, *Conus catus*.

The cat cone is abundant along the tropical shores of the Indian and Pacific Oceans and as far south as North West Cape in Western Australia. It hides in crevices in wave-swept intertidal reefs.

Potentially dangerous injury, toxic venom

For prevention, symptoms and treatment of injury see pages 146–147

Monk cone

Conus monachus Linnaeus, 1758

The shell of the monk cone, is up to 6 cm long and resembles the cat cone in both shape and texture. The colour pattern is variable but is usually of white and green or brown splashes of colour on a blue grey background. The white spiral ridges are usually patterned with fine spots of reddish brown. On the ridges in some specimens this brown colour is dominant, broken only by spots of white, while in others there is about the same amount of the two colours. *Conus achatinus* is a name which has been given to a variant of this species.

The monk cone has teeth shaped like those of other fish-eating cones and therefore is thought to eat fish. Although it has not yet been recorded as injuring humans, it is potentially dangerous, as experiments have shown that its venom can paralyse skeletal muscles of mammals.

The monk cone is quite common in Malaysian, Philippine and Indonesian waters. It is found along the northern coasts of Australia and south to about Jurien Bay on the west coast. Like most other cones it secretes itself in sheltered places during the day.

Potentially dangerous injury, toxic venom

For prevention, symptoms and treatment of injury see pages 146–147

Clay Bryce

▲ Monk cone, *Conus monachus.*

Striated cone

Conus striatus Linnaeus, 1758

The large shell of the striated cone, is rather thick and heavy and may be up to 10 cm long. The sides of its body whorl are curved, particularly near the flat spire. The shell is usually pale pink with an overlay pattern of large irregular blotches of purple brown. Darker, more solid blotches tend to form two broken irregular bands going around the central part of the body whorl. These blotches consist of fine close-set lines.

Clay Bryce

▲ Striated cone, *Conus striatus*.

Clay Bryce

▲ Striated cone, *Conus striatus* burrowing into sand.

The striated cone eats fish and should be regarded as potentially dangerous, though no serious injuries have yet been recorded. Tests have shown that its venom is less potent than that of the geographer cone, with at least some of the toxins being of a different type, paralysing skeletal muscle in a contracted state. Studies on the production of venom from its relatively small venom gland suggest that not enough could be produced to kill an adult human.

The striated cone is widespread in the Indian and Pacific Oceans but has not commonly been found in Western Australia. It is known to live around North West Cape, Onslow, Dampier and offshore reefs such as Rowley Shoals, where it hides in sand under coral slabs, boulders and ledges during the day.

Potentially dangerous injury, moderately toxic venom

For prevention, symptoms and treatment of injury see pages 146–147

Obscure cone

Conus obscurus Sowerby, 1833

The small obscure cone, looks like a miniature geographer cone, having a thin, smooth, almost cylindrical, violet shell of up to 4 cm in length, with a brown pattern of indefinite spiral bands. However, the shell has a spire which is relatively higher than that of the geographer cone and it does not have a triangular colour pattern. Inside, the shell is violet, clouded by the exterior colour pattern showing through.

Clay Bryce

▲ Obscure cone, *Conus obscurus*.

Although it feeds on fishes, the injuries reputedly caused by this species seem to have affected only the area around the site of the puncture wound and have not been severe or long lasting. It is possible that this small cone could not secrete enough venom to cause serious injury to humans but it should nevertheless be treated with caution.

The obscure cone shelters under coral, usually buried in sand. It lives in tropical areas of the Indian and West Pacific Oceans and, in Western Australia, has been found as far south as Barrow Island.

Localised injury, toxic venom
For prevention, symptoms and treatment of injury see pages 143–147

Textile or Cloth-of-gold cone

Conus textile Linnaeus, 1758

The textile cone, has a fairly smooth, glossy shell which is rather heavy for its size of up to 10 cm in length. Its shape varies from slender and fairly straight sided to squat with bulging sides, and its moderately high spire may have a straight or concave outline. The most notable feature of the shell is its tent colour pattern with triangular patches of the white background outlined in dark brown. Broken areas of tan, which are crossed by irregular dark brown radial lines, occur on the spire and form spiral bands running around both the body whorl and the spire. The interior of the shell is white.

During the day, textile cones hide under stones, coral slabs and ledges, and probably come out to feed at night. They feed on molluscs including

Clay Bryce

▲ Textile or Cloth-of-gold cone, *Conus textile*.

Clay Bryce

▲ Anterior end of Textile or Cloth-of-gold cone, *Conus textile.*

other cones, and can even kill an attacking octopus. Their venom does not appear to directly paralyse the skeletal muscle of mammals. Nevertheless, textile cones should be treated with great caution as there is evidence that their venom might affect the central nervous system.

The textile cone lives in tropical waters of the Indian and Pacific Oceans, and is fairly common on the Western Australian coast south to about Point Coates.

Potentially dangerous injury, toxic venom

For prevention, symptoms and treatment of injury see pages 143–147

Pearled cone

Conus omaria Hwass *in* Bruguière, 1792

The pearled cone resembles the textile cone. However, its shell is smaller, being less than 8 cm long, and it tends to be more slender with a lower spire and a more sharply-angled shoulder. The white background colour is broken into large well defined but irregular triangular areas. Large brown patches, delicately spotted with white, form two broken but wide spiral bands around the body whorl. The white shell interior is seen through the broad aperture.

The venom of this mollusc-eating cone has been recorded as having caused severe symptoms in humans and so, like the textile cone, should be handled carefully.

Clay Bryce

▲ Pearled cone, *Conus omaria.*

The pearled cone is found through tropical areas of the Indo-West Pacific region, including the north coast of Western Australia. It is rarely seen alive, although it lives in shallow coral reef areas as well as in deeper water.

Potentially dangerous injury, toxic venom

For prevention, symptoms and treatment of injury see pages 143–147

Anemone cone

Conus anemone Lamarck, 1810

The shell of the anemone cone, *Conus anemone*, is extremely variable in shape, pattern and colour. The soft parts are usually in shades of pink, and the background colour of the shell can range from almost white through yellow and pink to blue. The mottled pattern of red, brown or purple that overlies this background colour may be light or heavy, forming two or three broken and sometimes indistinct bands around the body whorl. The shell can be moderately heavy or lightweight and

Clay Bryce

▲ Anemone cone, *Conus anemone.* Very variable in shape, colour and pattern.

fragile-thinner shells usually being more rounded than thick shells. Shells that are apparently mature range from about 2.5 to 6.5 cm in length

Although the anemone cone is thought to be a worm eater and so would not be expected to be included among dangerous cones, it is known to have caused moderately painful but localised swelling of its victim's limbs (G. W. Kendrick & A. Brearley, personal comments). It is common along the coasts of the more highly populated areas of southern Australia.

The anemone cone occurs along the southern coast of Australia and along the western coast to just north of Geraldton in Western Australia. It is commonly found under rocks and sometimes in and on sand in intertidal areas and in deeper waters.

Localised injury, moderately toxic venom

For prevention, symptoms and treatment of injury see pages 143–147

Clay Bryce

▲ Anemone cone, *Conus anemone* — with a horse-shoe limpet (*Hipponix conicus*) attached to its spire.

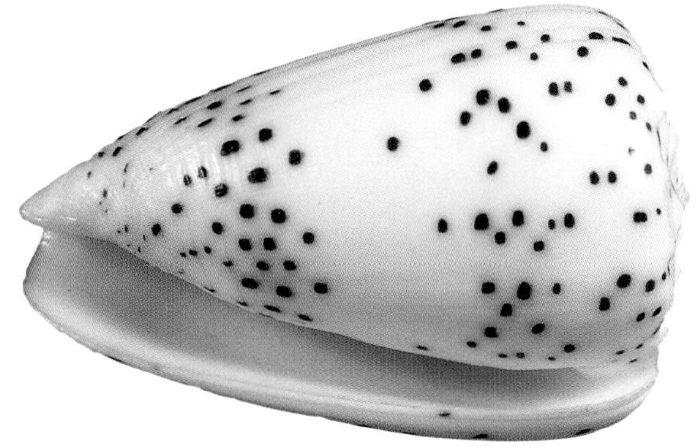

Clay Bryce

▲ Flea cone, *Conus pulicarius.*

Flea cone

Conus pulicarius Hwass *in* Bruguière 1792

The flea cone eats worms and has a fairly heavy cone-shaped shell with a low but pointed spire. The sides of its body whorl are slightly rounded anterior to its shoulder, which is knobbed. Black spots are scattered over the white shell and form two indistinct spiral bands where the spots are larger and denser than elsewhere. The interior of the shell, as seen through the fairly narrow aperture, is pale and usually white.

As with other worm-eating cones which may injure humans, the venom this cone injects causes bleeding and pain, localised around the site of the sting. The flea cone is common in the tropical Indo-West Pacific region and its distribution in Western Australia extends to an area south of North West Cape. It is usually found buried in sand, as are the worms on which it feeds.

Localised injury, moderately toxic venom

For prevention, symptoms and treatment of injury see pages 143–147

Victoria cone

Conus victoriae Reeve, 1843

The Victoria cone is a variable species, both in its shell shape and colour pattern. Most shells grow to about 7 cm in length and have a relatively high spire behind an angled shoulder. The sides of the body whorl are straight or slightly convex. The base colour of the shell is white, generally with radial streaks of a blue tinge. Overlying this is a tent pattern in brown or tan, with two or more darker spiral bands around the body whorl. Similar dark patches on the spire whorls are generally radially aligned. Those cones living in more southern waters have a more pinkish shell with a paler tan to orange colour pattern.

The Victoria cone lives on reefs and soft substrates in intertidal and shallow waters along the Western Australian coasts north of the Fremantle area, with its range extending into Northern Territory waters. It is generally abundant along the mainland coasts of Pilbara and Kimberley regions.

This cone has not formerly been considered particularly dangerous to humans, having been observed feeding on the gastropod *Cantharus erythrostomus* near Dampier, Western Australia (Kohn 2003) — having discharged 5 teeth into its victim. Nevertheless the components of the complex venom of this species have been found to be very active chemically and their effects and pharmacological potential are being investigated (see 'Painkiller extracted from cone venom', page 145).

We know of no records of humans having been stung by *Conus victoriae* and so the effects of envenomation are also unknown. It is thought to be related to *Conus textile* (Röckel, Korn & Kohn 1995) and so its venom might produce similar symptoms to those of that molluscivorous species.

Clay Bryce

▲ Victoria cone, *Conus victoriae*.

Octopus, Squid and Cuttlefish

Phylum Mollusca
Class Cephalopoda

Cephalopod molluscs, including octopus, squid, and cuttlefish, are regarded as the most highly evolved group of molluscs with a way of life more like that of fishes than of other groups of molluscs (Mangold, Clarke & Roper 1998). They are active predators with very efficient sense organs and complex nervous systems that allow them to react quickly. They locate their prey mainly by sight, having eyes which can see objects much as those seen by human eyes. In this way their eyes are unlike those of other molluscs such as cowrie or cone shells that can distinguish only between light and shade. The large cephalopod brain not only stores complex inherited behaviour patterns but can also learn and remember from its own experience.

Clay Bryce

▲ This solitary reef dwelling cuttlefish, *Sepia apama*, is curious and territorial, and may become aggressive to divers.

Like other groups of molluscs, the octopus, squids and cuttlefish have a soft but muscular body containing the body organs. The gills and the external openings of the digestive, excretory and reproductive systems are in the mantle cavity. However, these groups of cephalopods lack an outer protective shell. Eight arms, each with one or two rows of suckers, radiate from around their ventral mouths. Inside the mouth is a pair of jaws shaped like a parrot's beak, and a muscular 'tongue'. A membrane in which are embedded the bases of many rows of hard sharp teeth covers this 'tongue'. This toothed membrane, typical of cephalopods and many other molluscan groups, is called a radula.

In addition to their eight arms, squid and cuttlefish have a pair of longer tentacles that reach out to catch hold of their prey. A horny-toothed rim strengthens the edges of the suckers on both the arms and the tentacles of squid and cuttlefish. In some deep-water squid species the suckers on the tentacles and, more rarely, on the arms are replaced by long, sharp, horny hooks.

Most species of squid and cuttlefish can swim powerfully. Normally they use fins running along the sides of the mantles or nearer to the hind ends of their bodies. The muscles of the body and fins pull against an internal shell that, in squids, is slender horny, transparent and flexible but in cuttlefish is the wider rigid chalky cuttlebone often collected from beaches and fed to birds. Squid and cuttlefish may swim in schools or alone, and they hunt fishes, crustaceans and other cephalopods in the open ocean or in shallower coastal waters. When needing an extra burst of speed to catch prey or to escape from danger, these and other cephalopods use jet propulsion, forcing a stream of water from the mantle cavity out through a muscular and flexible mantle funnel.

Most cephalopods possess an ink gland from which they can puff out a cloud of dark pigment when disturbed. This cloud, appearing suddenly from the mantle funnel, startles a pursuer and momentarily hides the cephalopod, which instantly becomes paler and swims away.

Octopus lack the internal shell, the two long tentacles and the toothed suckers typical of squid and cuttlefish, and most of them do not have fins. Instead, they generally creep across rocks and sand and through weed searching for crabs or molluscs for food. On occasions, they do swim by jet propulsion but usually only for short periods as most tire quickly. When hunting at night or in dark caves, octopus probe with their eight muscular but sensitive arms, picking up the scent of their food. They do this by using sense organs located on the skin of their suckers that are also used to grasp and hold their prey.

Different octopus species vary in the methods they use to subdue and kill food animals. Some bore through the shell of well-protected molluscs using fine accessory teeth within the mouth. Others are reported to use their jaws to bite animals such as crabs and to inject saliva into the wound. But many species hold the victim within an 'umbrella', formed by their webbed arms. Saliva is squirted into this confined space and is absorbed into the tissues of the victim, paralysing and killing it.

Octopus have two sets of salivary glands opening into the mouth. One pair in the 'neck' region behind the eyes produces thick saliva which may

Clay Bryce

▲ Most octopus creep over the substrate, changing their colour pattern and sometimes their skin texture to merge into the background.

contain toxic substances. These glands are particularly large in the blue-ringed octopus. The secretions from a smaller pair of glands may contain a spreading agent to dilute the venom and perhaps to aid its penetration. The saliva of many octopus species is digestive in function and loosens the attachment of the crab's or mollusc's muscles to its shell. The octopus then pulls the partly digested flesh from even the finest cavities so that a clean but almost unbroken shell remains. The jaws may bite off pieces of the prey's body, which are drawn in by the tooth-covered 'tongue', or radula.

Octopus escape from enemies such as large fish and seals by jet-propelled swimming. They, and other cephalopods, can also change their skin texture and colour pattern for camouflage. Muscles control the size of the pigment cells in the skin so that, as these muscles contract and expand, the skin colour darkens or lightens. The pigment cells are of different colours, and so cephalopods may also change their colour.

FIGURE 12:

The organs of a blue-ringed octopus used in prey capture and feeding.

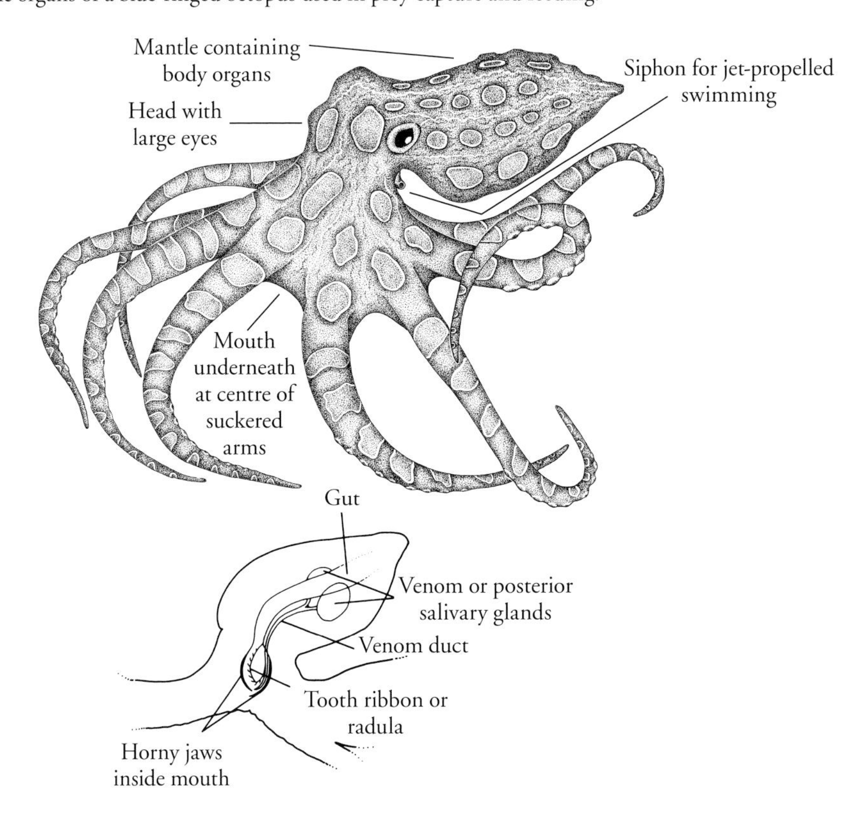

When threatened or when excited by the presence of prey or sometimes of another cephalopod, the colour change can be dramatically swift and repetitious. Many colour patterns are characteristic of a particular species, although some types of patterns are more general and may serve for inter-species communication (Messenger 2001).

CEPHALOPOD BITES – EXCEPT FROM THE BLUE-RINGED OCTOPUS

The bites of squid, cuttlefish and octopus produce symptoms that are related to the amounts and types of chemicals present in their saliva. In most species that have been investigated, these components consist of digestive enzymes and toxins. A wound through which these substances have entered might bleed freely, although usually victims have reported little associated pain.

Cephalopod's suckers which cling might produce discomfort but cause no harm except, perhaps, for a very slight abrasion from the serrated sucker rings of a squid or cuttlefish. However, panic can be much more dangerous to a diver than the bite of any cephalopod — even that of a venomous species — could ever be. If a cuttlefish should grasp a diver's face mask, attracted by its shining glass and metal, its presence should be disregarded as far as is possible and the rules governing a safe ascent to the surface should be rigidly followed. The knowledge that excited, frightened or injured cephalopods will spurt a cloud of 'ink' and grasp onto anything within reach should help a diver to avoid the panic reaction which might otherwise lead to the 'bends' or drowning.

The symptoms caused by cephalopod bites, other than those of the blue-ringed octopus, usually pass off after a few hours and the slight wounds should heal without any complication.

INJURY PREVENTION OF CEPHALOPOD BITES EXCEPT FOR THOSE OF THE BLUE-RINGED OCTOPUS

- If 'attacked', grasp the cephalopod's body around its 'neck' above the ring of arms and below the eyes.
- Remember that adhering suckers cause no harm except, perhaps, a very slight abrasion from a squid's or cuttlefish's serrated sucker rings.
- Avoid panic underwater, which could lead to the 'bends' or drowning.

SYMPTOMS OF CEPHALOPOD BITES EXCEPT FOR THOSE OF BLUE-RINGED OCTOPUS

- Wounds might bleed freely, though there is usually little pain.
- Redness and swelling around the wound may develop, even some hours later.
- Some tingling, numbness, a feeling of light headedness and weakness may occur.

TREATMENT OF CEPHALOPOD BITES EXCEPT FOR THOSE OF BLUE-RINGED OCTOPUS

- Symptoms usually pass off after a few hours.
- Clean wound and cover it.
- Seek medical attention for large wounds which may require stitches or other surgical action.
- Tetanus booster may also be necessary.
- Wounds usually heal without any complication.

For prevention, symptoms and treatment of injury from the blue-ringed octopus, see page 167–168

THE 'BENDS' OR DECOMPRESSION SICKNESS

The bends can occur in a scuba diver who is breathing compressed air from a tank and who ascends too quickly. The problem is that nitrogen which has dissolved in the blood at depth comes out of solution as bubbles when the pressure decreases rapidly. The bubbles can form anywhere — joints, heart, gut, brain — and be very painful, cause long lasting injury or even be life-threatening.

The immediate treatment requires rapid recompression. This may involve descending to the previous water depth in order to re-dissolve the nitrogen, and then coming up slowly and in stages. If this cannot be done, the only alternative is to simulate that process by the use of a hyperbaric chamber, found only in specialist hospitals such as Fremantle Hospital, Western Australia. In all cases the immediate administration of oxygen is required.

BLUE-RINGED OCTOPUS

Phylum Mollusca
Class Cephalopoda
Family Octopodidae

The various species of blue-ringed octopus are grouped within the genus *Hapalochlaena*. They are dealt with separately here from species of the genus *Octopus* and other cephalopods because of the much greater toxicity in their saliva which can cause human death.

The most obvious distinguishing characteristic of a blue-ringed octopus is the colour pattern of its skin. On a background of mottled brown are rings and/or streaks of iridescent blue which change in size and intensity

Clay Bryce

▲ Except when excited, blue-ringed octopus are not colourful but blend with the colour of their background.

depending on the animal's level of excitement. When the octopus is close to prey or when it feels threatened, these rings or bands enlarge dramatically and become more vivid, their colour highlighted by an expansion and a deepening of the colour of the dark areas beneath them. This colour pattern appears rapidly and can change just as quickly back to the camouflage pattern of various shades of mottled yellow-brown with the blue bands or rings becoming almost invisible.

Blue-ringed octopus are also superficially distinguished by being generally smaller, having relatively shorter arms and having a web of skin between the arms that is relatively wider than in most types of octopods. Although they are said to have ink glands for only a short period after hatching, a full-sized specimen from Broome in Western Australia did put out a small quantity of brown ink when extremely irritated.

At least five species of blue-ringed octopus are known to live in the shallow waters along the Australian coastline (Norman 2000) and three of these have been recorded from Western Australia.

Some, if not all, of the Australian species of blue-ringed octopus use highly toxic venom in their saliva to quickly kill their prey, which generally consists of small crustaceans such as crabs. This venomous saliva has caused human deaths.

The first indication that the bite of a blue-ringed octopus could be fatal occurred in the 1950s with a death in the Northern Territory. Another occurred in New South Wales in the 1960s (Norman & Reid 2000). Recent research has shown that the saliva of a blue-ringed octopus can be more potent than the venom of the most deadly snake. Further, the toxins contained in the saliva are actually produced by bacteria which live in the salivary glands of the octopus (Norman 2000).

There is some doubt as to whether blue-ringed octopus always inject the venom into their victims, be they crabs, fish or humans. Some reports indicate that no incisions could be found on the skin of the victims of fatal and non-fatal attacks, but only faint bruising which faded rapidly. On the other hand, the skin of other victims was lacerated to a depth of 5 mm, probably by the beak of the octopus.

The venom of some species of the blue-ringed octopus has been intensively investigated and has been found to contain two main toxic elements, maculotoxin and hapalotoxin.

Maculotoxin (said by some workers to be identical with the powerful nerve toxin, tetrodotoxin, found in puffer fish) is apparently not toxic to crabs and other crustaceans but is toxic to fishes. It may be used in defence as well as in predatory actions upon fish. The other toxin, hapalotoxin, is effective on crabs and is similar to some less toxic substances produced by other crustacean-eating cephalopods.

Both maculotoxin and hapalotoxin affect the transmission of impulses between nerves and skeletal muscles in mammals. Their relatively small molecules do not produce an antigen response, so complicating the development of an antivenom.

INJURY PREVENTION OF BITES FROM BLUE-RINGED OCTOPUS

- Do not catch or hold a blue-ringed octopus, or allow one to crawl on the skin. They are more likely to bite if irritated by being lifted out of the water and handled roughly.
- Because these small octopus may inhabit empty bottles, shells (even quite small ones), etc., do not carry any such objects which have been collected underwater in the hand, in a pocket or tucked inside a wetsuit as they may contain an undetected octopus.
- Be careful when collecting or sorting mussels, as blue-ringed octopus often live among them.
- When hauling crab or prawn nets or rock lobster pots remember that blue-ringed octopus often shelter in them and may become upset when disturbed.

SYMPTOMS OF BITES FROM BLUE-RINGED OCTOPUS

- Respiratory failure is a real danger, indicated by the victim's breathing becoming shallow and the skin colour turning blue.
- The bite of a blue-ringed octopus is rarely felt and does not cause pain or swelling, though some bleeding may occur.
- The bite may be so small that it cannot be located; the first indication of having been bitten, occurring within minutes, is of numbness in the area of the bite and then in the lips and tongue.
- Increasing difficulty with speaking and swallowing occurs as the muscles of the throat and larynx become paralysed.
- General paralysis may develop rapidly.

TREATMENT OF BITES FROM BLUE-RINGED OCTOPUS

- Victim is likely to vomit and so, if the victim is not required to be laid on the back for EAR, position in the recovery position (lying on one side) to minimise the danger of choking (see recovery position, pages 224–226).
- If breathing starts to fade and victim turns blue, start expired air resuscitation (EAR) and maintain until arrival at a medical centre that can take over ventilation (see EAR, pages 227–228). Life support may be needed for up to 24 hours.
- Vomit may lodge at the back of the throat of a paralysed victim, causing choking. Constant checking of the airway to the lungs is essential. Clear any obstruction immediately, using fingers if necessary.
- Immobilise the whole body as far as possible.
- If the site of the bite is known, apply pressure with a pressure bandage around the affected limb or with a pressure pad and firm bandage if bite is elsewhere, as for snakebite.
- If the victim's eyes cannot move, shade them to protect retinas from the sun.
- Never leave the victim unattended as EAR might become necessary. Although totally paralysed, a victim might still be conscious and able to hear but not respond. Explain all that is happening to the victim, give reassurance that help is coming and do all possible to ease distress and avoid causing panic.
- Taking care of the victim is demanding; it requires constant watchfulness and must persist until medical help is at hand or until the effects of the bite have worn off.

Blue-ringed octopus of southern Western Australia

Hapalochlaena species

Phylum Mollusca
Class Cephalopoda
Family Octopodidae

The small blue-ringed octopus, *Hapalochlaena maculosa* (Hoyle, 1883), is common around the south-eastern coasts of Australia, and is held to be responsible for a fatality in the Sydney area in 1967.

A very similar but currently unnamed species of blue-ringed octopus inhabits the south-western coasts of Western Australia. However, it and other small *Hapalochlaena* species are not often noticed because of their small size (rarely exceeding 12 cm across their arm span), their powers of camouflage and the speed with which they can move from one place of

Clay Bryce

▲ *Hapalochlaena* sp. An un-named blue-ringed octopus from the southern coasts of Western Australia, similar to the south-eastern Australian *H. maculosa*, flashes its brilliant blue rings and bands when disturbed.

concealment to another. They are found sheltering within dead gastropod shells, clumps of mussels and discarded bottles and cans. They frequent algal-covered boulder piles exposed to pounding waves as well as sheltered seagrass beds, from the intertidal areas to depths of at least ten metres.

The females of this species often inhabit empty, hinged bivalve shells such as razor clams when laying and brooding their eggs. Their suckered arms hold the shell valves together to protect themselves and the eggs they guard. These eggs are relatively large for such a small octopus and are arranged in clusters attached to one another but not to the substrate. About 100–150 egg capsules, each approximately 7 mm long, may be held in the groove between the mother's body and the webbed base of her uplifted arms. It is thought that as in most cephalopods studied, the females do not feed while brooding their eggs, which take about two months to develop and hatch.

The young blue-ringed octopus of the species *H. maculosa* are able to crawl on the seabed almost immediately after hatching, staying with their mother for several days before moving off to fend for themselves. Still not eating, the mother might live for a few more weeks before dying at the age of about one year. It is presumed that the male of this species also dies after breeding, as happens in some of the few other species of octopus that have been studied.

Very little is known about the habits of the Western Australian species but they are thought to be similar to south-eastern Australian species. Newly-hatched *H. maculosa* are anti-social to the point of eating each other but, by the time they are one month old, can catch and kill small crabs which seem to be their preferred food. Their toxin-producing salivary glands appear to function from the time they hatch from the egg (Stranks & Lu 1991).

Bathers and fishers in southern Western Australia have been bitten by blue-ringed octopus but have survived, presumably mainly due to an increased awareness and a quick application of appropriate first aid measures.

For prevention, symptoms and treatment of injury from the blue-ringed octopus, see pages 167–168

Greater blue-ringed octopus

Hapalochlaena lunulata (Quoy and Gaimard, 1832)

Phylum Mollusca
Class Cephalopoda
Family Octopodidae

Hapalochlaena lunulata (Norman & Reid 2000) is the larger of the two inshore species of blue-ringed octopus living along the northern coasts of Western Australia and eastwards from there. It is nearly twice the size of the southern and the other north-western blue-ringed octopus species, growing to about 15–20 cm when measured across its extended arms. Its skin is usually quite smooth and of a mid-tan colour, and its blue rings are large, fine and rarely backed by black areas.

This blue-ringed octopus has been found south of the Houtman Abrolhos, living under rocks and dead coral on reef flats and also in tidal pools in more muddy areas. Like related species, it is not often seen because of its retiring habits and ability to camouflage.

A human death caused by the bite of a blue-ringed octopus occurred near Darwin in the Northern Territory in the 1950s. Only a very small

Clay Bryce

▲ *Hapalochlaena lunulata* from the Dampier Archipelago, Western Australia (Specimen identified by Mark Norman).

▲ The greater blue-ringed octopus from north-western Australia can be recognised by its large open rings and smoother skin.

puncture wound was found on the victim's skin and there was no development of localised swelling or pain around the bite. This fatality has been attributed to the species *H. lunulata*.

For prevention, symptoms and treatment of injury from the blue-ringed octopus, see pages 167–168

A smaller northern blue-ringed octopus

Phylum Mollusca
Class Cephalopoda
Family Octopodidae

An apparently un-named species of blue-ringed octopus lives along the northern coasts of Western Australia. It is similar in shape and size to the species of more temperate Australian waters but has a slightly different arrangement of the blue rings on its body. This northern Australian species inhabits shallow intertidal pools as well as turbid deeper waters around reefs and jetty piles.

Although nothing is known of the effects of this species' bite on humans, an aquarium specimen killed a large sea snake within a minute of leaping onto its back. This could indicate that the venom of at least this species of blue-ringed octopus is used as a defence against natural predators as well as for killing prey.

For prevention, symptoms and treatment of injury from the blue-ringed octopus, see pages 167–168

Poisonous Molluscs

Most molluscs are edible and many are widely used for food, being well flavoured and nourishing. Some, such as oysters, may even be regarded as luxuries. Unfortunately, some people develop allergies to one or more types of molluscan foods, which others can eat without problems (see Allergies, pages 219–223).

Some molluscs may be occasionally contaminated by bacteria, by chemical pollutants or by naturally occurring poisons and a very few molluscan species contain poisons which they produce themselves. Sometimes the symptoms from these different causes can be similar and so may be easily confused.

Apart from allergic reactions, illness caused by eating molluscs is fortunately uncommon in Australia.

Clay Bryce

▲ Many nudibranchs, such as this attractive *Phyllidia coelestis* Bergh 1869 from Scott Reef Western Australia, are distasteful and even poisonous to fishes, some other invertebrates and probably humans.

POISONS AND INFECTIONS FROM POLLUTION

Most farming or commercial harvesting of molluscs such as mussels, oysters or scallops is carried out in shallow coastal waters, usually in sheltered embayments and estuaries where the levels of planktonic food, and so the growth rates of the molluscs, are highest and where rough seas rarely interfere with the fisheries tasks. In such sheltered areas recreational fishers also gather these clams, and cockles and other molluscs for their own use. Unfortunately, such shallow coastal waters are sometimes polluted by domestic and industrial sewage, by effluents from shipping harbours or marinas, or by drainage from the adjacent countryside. The same threats may still apply, even if moderated, in some less obvious areas.

Molluscs taken from such areas, particularly filter feeding bivalves, have caused illness. Attached to farming gear, rocks or jetty piles, living on

Clay Bryce

▲ Some people may develop allergies after eating molluscs such as these blue mussels from southern waters.

Clay Bryce

▲ Many if not most species of bivalve molluscs, including this razor clam *Pinna bicolor*, may concentrate pollutants in their tissues as they filter water for food and oxygen.

the surface of or burrowing under soft substrates, most bivalves feed by filtering minute living planktonic organisms or particles of dead animal and plant tissue from the water passing through their perforated gills. Particles suitable for food are passed along into the mouth while rejected material is tangled into mucus balls and flushed out of the bivalve's body.

This feeding system can result in the concentration, though sometimes only temporarily, of any disease causing micro-organisms, heavy metals and organic pollutants in the body cavities or within the actual tissues of the bivalves. Such concentrations may also occur in some gastropods, such as periwinkles or creepers, even though they feed upon encrusting or leafy algae or upon organic debris.

Although most infective micro-organisms usually do not survive for long in the sea or even inside molluscs, pollutants (including organic chemicals such as herbicides, and inorganic chemicals such as zinc and other heavy metals) may be even further concentrated in molluscivorous molluscs further up the food chain. If consumed in sufficient quantity, these pollutants may be even further concentrated in the human body, sometimes to dangerous levels.

POISONING AND INFECTION PREVENTION

Molluscs from areas likely to receive inadequately treated sewage or run-off from pastures, service stations, factories and similar industries should not be eaten.

Molluscs from shallow waters frequented by seagulls should also be avoided, as these birds can carry micro-organisms which cause disease. Unfortunately, many such areas are most favourable for the growth of molluscs, and so for oyster and other farming and for fishing activities.

Holding suspect molluscs for 24 hours or more in clean salt water from a safe area can reduce the danger from micro organisms but not necessarily from chemical pollutants.

POISONS MADE BY MOLLUSCS OR BY THE FOOD ORGANISMS THEY EAT

Relatively little is known of the chemical nature, origin or formation of most of the poisons which protect some and perhaps many of the soft, fleshy and often brilliantly coloured nudibranchs, sea hares and other groups of thin-shelled or shell-less gastropods.

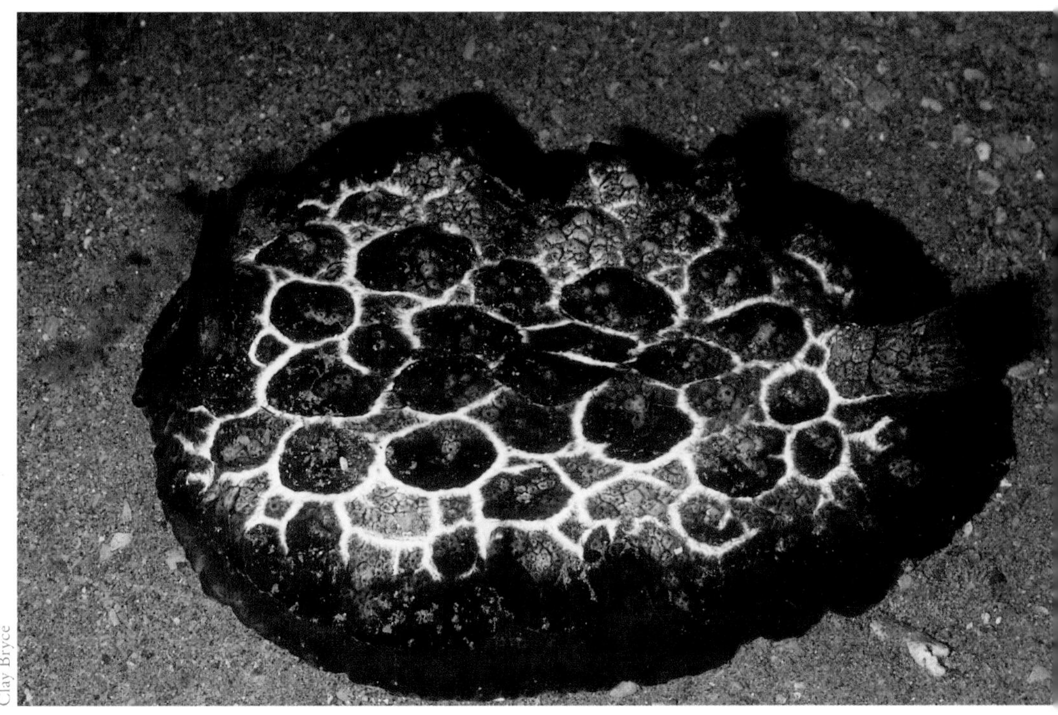
Clay Bryce

▲ *Pleurobranchus forskalii* from the Dampier Archipelago.

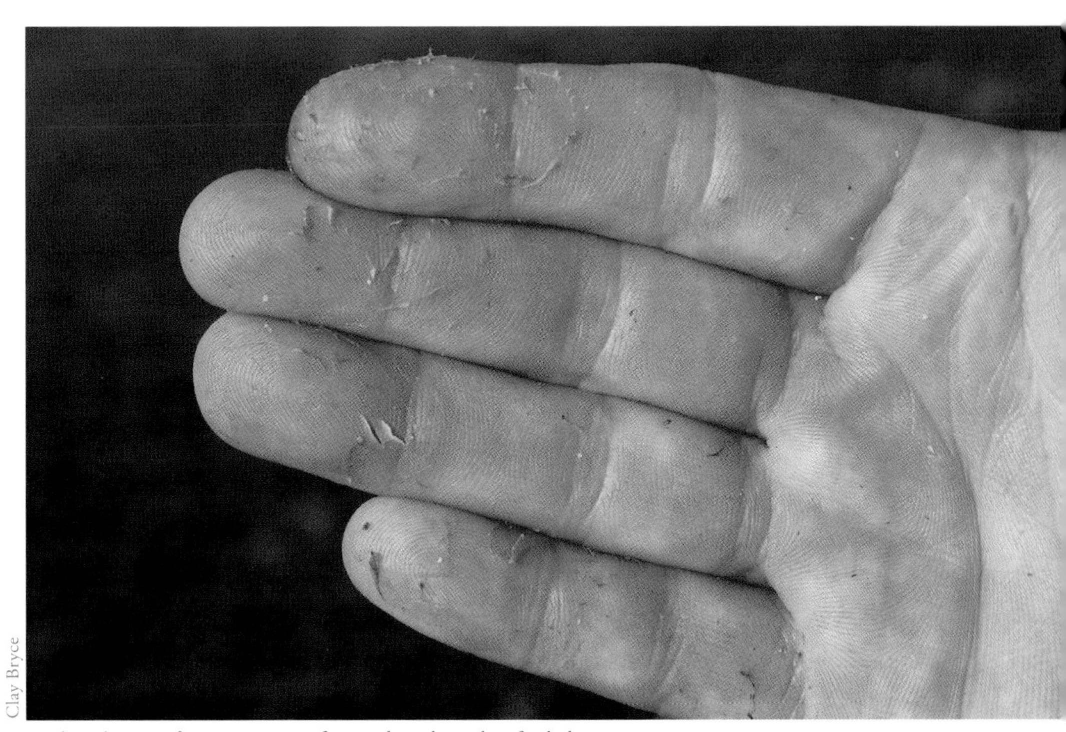
Clay Bryce

▲ Skin damage from secretions from *Pleurobranchus forskalii*.

The sea slug, *Pleurobranchus forskalii* (see page 177) secretes from its skin a mucus containing a substance which can damage the skin of anyone handling it. It is probably distasteful if not poisonous to natural predators.

Some nudibranchs, such as some species of the large and diverse families Chromodorididae and Phyllidiidae (e.g. *Phyllidia coelestus* page 173), are known to store certain distasteful and sometimes poisonous chemicals derived from the sponges they eat. These chemicals, whether from the sponges or from the nudibranchs, appear to have the same effect in deterring would-be predators.

Similarly, at least some species of sea hares of the family Aplysiidae derive poisons from their food. A species of *Stylocheilus* that feeds on organisms such as filamentous cyanobacteria may contain toxins acquired from that source. Some of the secretions from glands in the skin and other body organs of these sea hares have been shown to have a defensive function, making the soft bodied molluscs unpalatable and perhaps even poisonous to predators.

Research into some components of the defensive purple fluid secreted by the Australian Indo-West Pacific sea hare, *Aplysia dactylomela*, has shown that they exhibit antibacterial and blood coagulating characteristics which may prove to be of medical interest (Melo *et al.* 2000).

More is known about paralytic shellfish poison (PSP). The toxic substances included in this group of poisons, such as saxitoxin, gonyautoxin and tetrodotoxin, are derived from substances produced by algae and absorbed and concentrated by molluscs. Somewhat similar toxins have been given the names neurotoxic shellfish poison, diarrhoeal shellfish poison and amnesic shellfish poison — their names indicating their effect on human victims i.e. affecting the nervous system and causing paralysis, causing diarrhoea, and causing error in and/or loss of memory. Molluscs absorbing and retaining these poisons might not be affected themselves and may, in fact, derive some protection from predators due to the accumulated and bitter tasting toxins.

One of the neurotoxins which may accumulate in molluscs causes paralysis. It is very like, if not identical with, the tetrodotoxin contained in certain organs of puffer fish belonging to the family Tetraodontidae. Its small molecules are not affected by the digestive process of humans and, once absorbed into the bloodstream, affect the nervous system causing a loss of sensation and of controlled movement, along with general body pain. If skeletal muscles are sufficiently paralysed, breathing is affected and can even cease, causing death.

Some such toxins are initially produced by minute single-celled organisms called dinoflagellates which may be either floating at the mercy of oceanic currents (planktonic) or living on the substrate or on other organisms growing on it (benthic).

Planktonic dinoflagellates such as species of the genera *Gonyaulax* and *Gymnodinium* inhabit both temperate and tropical waters and, under certain oceanic conditions, can grow and multiply very quickly to form the very high densities called algal blooms. These blooms may colour the sea and may cause it to glow at night. Some give the water a red colour and waters affected by such algal blooms are called red tides. It is at such times that the level of toxins in bivalves feeding on these dinoflagellates may become very high. In some areas, human health problems involving the respiratory tract have even been attributed to the transport inland of such dinoflagellates from coastal red tides by strong onshore winds.

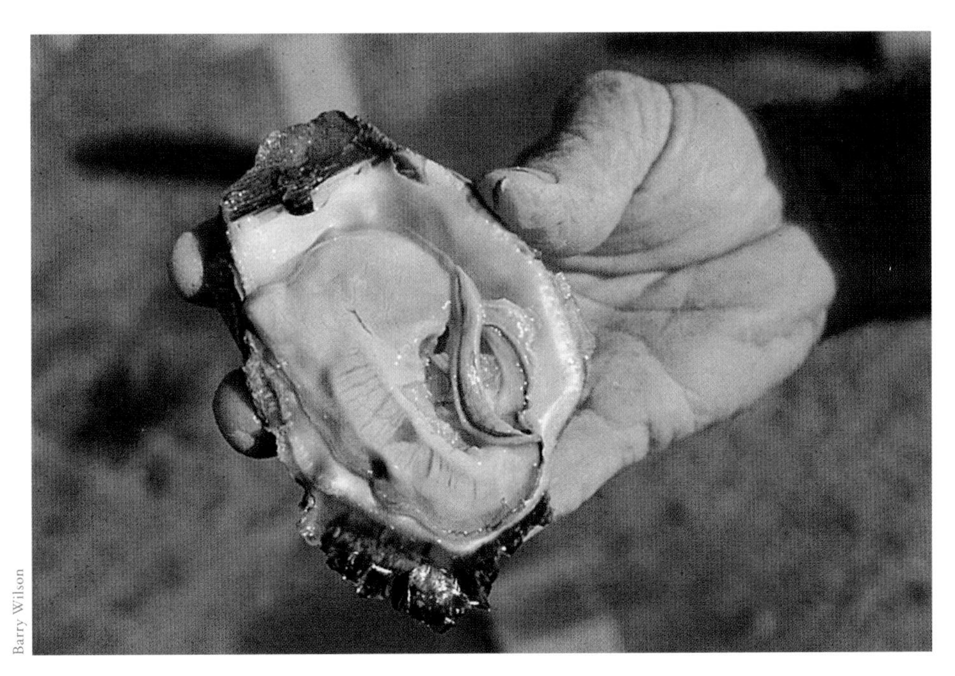

Barry Wilson

▲ The monitoring of paralytic shellfish poison (PSP) levels in oysters such as this large black-lip oyster, *Saccostrea echinata*, has become necessary as oyster farming, particularly in northern Australian waters, develops.

Some dinoflagellates, including some which contain toxins, are reported as having been introduced into Australian waters through the discharge of ships' ballast water. Such an introduction, followed by an algal bloom, has caused the closure of Tasmanian shellfisheries a number of times since the mid-1980s due to the high toxicity levels in the shellfish (Jones 1991). The incidence of such problems appears to be increasing worldwide, with grave implications for aquacultural industries (Hallegraeff 1993).

High levels of similar poisons occur, apparently spasmodically, in some weed-eating herbivorous gastropods such as some turban and conch shells inhabiting tropical coral reefs, and in some carnivorous gastropods such as whelks and triton shells of more temperate waters. Dinoflagellates cannot be the immediate origin of these toxins as these molluscs are not themselves filter feeders. The toxic chemicals from dinoflagellates may enter this food chain by the carnivores' predation on filter feeders such as bivalves, sponges, etc. or because of herbivores browsing on plants on which the dinoflagellates have settled. Apparently the toxins accumulate in the gastropods' tissues, mainly in their digestive glands.

If toxin-containing molluscs are eaten by humans or fishes, symptoms of paralytic shellfish poisoning (PSP) can develop, either rapidly or after a delay of a few hours. The severity of the symptoms depends on the amount of poison eaten and on the amount of other food in the stomach at the time. Different but often closely related species of bivalves have different abilities to concentrate and store these toxins. Much research work has been done and continues in many institutions throughout the world. The search for preventatives, assay methods and treatments is encouraged not only by considerations of human health but also by commercial concerns relating to a wide diversity of aquaculture projects.

Another aspect of research into such toxins is that concerned with the discovery of new drugs useful in medicine and other areas. The very fact that chemicals such as these are so active that they can cause interference with the passage of impulses along nerve pathways seem to indicate profitable areas of research. These might be in the area of 'pure' research on the actual functioning of the nervous system, in the area of medicine concerned with the treatment of neurological diseases or in the pharmaceutical area in the search for new anaesthetics, for example (see Wang *et al.* 2003).

POISONING PREVENTION

It seems that the muscular parts of gastropods affected by red tide or other poisoning may generally be eaten without ill effects if the gills and visceral mass containing the digestive and other organs are first cleanly removed.

No portion of similarly affected bivalves such as oysters, cockles or clams should be eaten because the digestive gland is widely spread through the tissues and because other organs of the bivalve may also be used for toxin storage.

Such toxins are soluble in water and affected by heat, and so mincing, repeated washing and cooking (particularly in a canning process) may reduce the toxin to tolerable concentrations.

Clay Bryce

▲ Tropical intertidal rocky reefs, such as this at North West Cape in Western Australia, may be covered with tightly packed coral rock oysters, *Saccostrea cucculata*. Some oyster species may retain PSP long after oysters of other species nearby have eliminated it.

Unfortunately, neither the presence nor the concentration of paralytic shellfish toxin is detectable by any change in colour or odour of the affected molluscs.

SYMPTOMS AND TREATMENT

- Should numbness, difficulty with speech and vision, muscular weakness, increased salivation, and diarrhoea occur in a person after eating molluscs, then vomiting should be induced immediately in all who have eaten the same food.
- However, do not induce vomiting in victims if paralysis has developed as there is a danger of vomit blocking the air passages. It might also be too late for vomiting to have any benefit.
- If the eyes cannot move or close, shade them to protect the retinas from the sun.
- If breathing becomes difficult, use expired air resuscitation (EAR)and maintain until breathing is normal again or until a medical station is reached (see EAR, pages 227–228).
- Victim should never be left alone because of the danger of respiratory failure.

CASE REPORT 1

Early in 2002, a number of dogs became ill after swallowing or even only mouthing dried sea hares of the large Western Australian sea hare species, *Aplysia gigantea* (Sowerby, 1869), which they had found in the drift on beaches around Geraldton WA. The dogs showed symptoms similar to those which would result from strychnine poisoning and treatment appropriate for that type of poisoning enabled a veterinarian to save at least some of the dogs.

(P. Taylor, personal comment, March 2002)

Similar toxins have caused human poisonings in Hawaii (Anon 1995), and also in Guam (Haddock 1993) and Japan (Noguchi, Matsui, & Miyazawa 1994), with some fatal results. The toxins involved in these poisonings were present in or on species of red algae, which are commonly eaten in those and other Pacific Ocean countries. However, it is thought that the toxins might have been present, not in the alga itself but in blue-green algae adhering to it.

However, an aplysiid species belonging to the genus *Dolabella* and even the tangled jelly-like strings of its egg masses are regularly eaten in the Philippines and some Pacific Islands

(Willan, personal comment, 2006)

Research into some components of the defensive purple 'ink' secreted by the Australian/Indo-West Pacific sea hare, *Aplysia dactylomela*, has shown that they exhibit antibacterial and blood coagulating characteristics which may prove to be of medical interest. (Melo *et al.* 2000)

CASE REPORT 2

In 1974, a single oyster eaten from the Great Barrier Reef caused moderately severe paralytic shellfish poisoning when eaten by one of the authors, Shirley Slack-Smith. However, another oyster of a different species, taken at the same time less than 30 cm away, caused no ill effects when eaten by co-author Loisette Marsh.

Echinoderms

Phylum Echinodermata

Echinoderms are diverse in appearance, ranging from stars to balls or sausage shapes, but all have a spiny skin (hence the name which means just that) under which is a hard skeleton. The skeleton consists of jointed arms in the feather stars and brittle stars, loosely joined plates in starfish, close fitting plates in sea urchins and minute rods, or spicules, embedded in the skin of sea cucumbers. Most echinoderms are slow moving, with a unique, water circulation system which operates rows of tube feet used for movement.

Humans use some sea urchins and sea cucumbers for food but many echinoderms contain toxic substances and others have venom glands associated with spines or with special pincer organs called pedicellariae.

Toxic saponins (sulphated steroidal glycosides) are present in starfish (Asteroidea) and sea cucumbers (Holothuroidea) but not in feather stars (Crinoidea), brittle stars (Ophiuroidea) or sea urchins (Echinoidea). These substances cause soap-like frothing of water, and make the animals extremely distasteful to predators. Some feather stars, brittle stars and sea urchins secrete a toxic mucus.

The spines of most starfish and sea urchins are non-venomous although their penetration results in a painful wound. If spines cannot be easily removed they will either be absorbed harmlessly or work their way out in time. Exceptions to this are crown-of-thorns starfish and sea urchin spines of *Echinothrix* species which must be surgically removed to prevent later

Clay Bryce

◀ Needle-spined urchin *Diadema setosum* in a sponge garden habitat off Dampier.

complications. Symptoms and treatment of wounds from the venomous spines of these echinoderms or from pedicellariae of other sea urchins are discussed under the species.

SYMPTOMS OF ECHINODERM WOUNDS

- Initial, sharp, stabbing pain.
- Pain may last for half an hour or much longer.
- Swelling and inflammation may occur.
- Fragments of spines will be visible in the wound or under the skin.
- If possible, collect a piece of the offending animal for identification in case wound is from a venomous species.

TREATING ECHINODERM WOUNDS

- Medical treatment should be sought for all but the most minor injuries, especially if any spine fragments remain embedded. Medical action may include surgery, tetanus booster injection to prevent development of tetanus, and prescription of general antibiotics and painkillers. The steps below can be followed for treatment of minor wounds, and for first aid.
- If possible, remove any obvious pieces of spine. Remove cleanly by pulling straight out, not by 'jiggling' which will cause the tip to break off (Edmonds 1981).
- Soaking injured part in vinegar may dissolve the spine fragments if they are close enough to the skin surface.
- Thoroughly clean and disinfect the wound to prevent secondary infection.
- Embedded spine fragments from non-venomous species will either be absorbed or work their way out over time.
- Injury may be complicated by a reaction to detached pieces of spine or to the skin covering the spine.
- Oral aspirin or paracetamol may be used for pain relief.

Refer to Infections and Allergies, page 219–223.

Feather Stars

Phylum Echinodermata
Class Crinoidea

Feather stars cling to rocks or coral by hooks on their underside. They feed by trapping small organisms with slender tube feet on the side branches (pinnules) of the arms. Food is passed down the arms in mucous strings to the mouth.

Feather stars are not known to be harmful to man but it is thought that some feather stars secrete toxic mucus.

Barry Wilson

▲ Feather star. The tube feet on the arms produce mucus (which may be toxic) to trap food particles, which are then passed to the central mouth.

Brittle Stars

Phylum Echinodermata
Class Ophiuroidea
Family Ophiocomidae

Ophiomastix annulosa (below) is one of the few brittle stars known to be toxic. It produces toxic mucus which causes paralysis and death in small animals, and so should be handled with care. It is common on tropical Indo Pacific reefs including the atolls off the north western coast of Australia.

In the same family, *Ophiocoma erinaceus* and *O. scolopendrina* are also recorded as being toxic, both are found on Australian coral reefs.

INJURY PREVENTION

- Although feather stars and brittle stars pose little threat to humans it would be wise for persons with sensitive skin to wear gloves if handling them.

TREATMENT

- If inflammation results from touching feather stars or brittle stars, apply antihistamine cream.

Clay Bryce

▲ One of the few brittle stars known to produce a toxic secretion is a common coral reef species, *Ophiomastix annulosa*.

Starfish or Sea stars

Phylum Echinodermata
Class Asteroidea

Many starfish (also known as sea stars) contain toxins in their digestive and reproductive organs which are not only poisonous when eaten but the poison can diffuse from living starfish into the water of an aquarium, killing molluscs, crustaceans and fishes.

Toxic water soluble substances called saponins have been isolated from members of the families Astropectinidae (*Astropecten indicus, A. polyacanthus*), Asterinidae (*Aquilonastra burtoni, Meridiastra calcar*), Acanthasteridae (*Acanthaster planci*), Oreasteridae (*Goniodiscaster scaber*) and Asteriidae (*Asterias amurensis*) but probably occur in other families that have not yet been tested. Starfish saponins have been analysed in a few species: major components of the toxin in a pest asteriid species (*Asterias amurensis*), accidentally introduced into Australia, are asterosaponin A and B. These are toxic to fishes, lethal to fly maggots and earthworms

Barry Hutchins

▲ If handled, the starfish *Plectaster decanus* may cause a rash lasting for several days.

and have caused vomiting in cats. The toxin in *Meridiastra calcar*, named calcarsaponin, is a haemolytic (cause rupturing of red blood cells) saponin. The highly surface-acting properties of starfish saponins make them haemolytic, anticoagulant and toxic to fishes in aquaria.

The tissues of *Astropecten polyacanthus* also contain the puffer fish poison tetrodotoxin and it has been shown that this species is attracted to bait containing tetrodotoxin (Saito and Kishimoto 2003).

Only two families of starfishes, Echinasteridae and Acanthasteridae, are known to have venomous spines. Echinasterids have a small body with long slender arms bearing spines on the net-like skeleton. Some species of *Echinaster* (an echinasterid) have thorny spines and surface pits from which a poisonous fluid is secreted. None of the six Australian species of *Echinaster* has caused injuries but *Plectaster decanus*, of the same family (see page 189), found on the southern Australian coast from New South Wales to Fremantle in Western Australia, has been recorded as causing contact dermatitis (Pope 1968).

CROWN-OF-THORNS STARFISH

Acanthaster planci (Linnaeus, 1758)

Phylum Echinodermata
Class Asteroidea
Family Acanthasteridae

The crown-of-thorns starfish, *Acanthaster planci* (see page 191), usually rare, has become notorious for its destruction of living coral over wide areas of tropical Indo-Pacific and Australian coral reefs when in plague numbers. It feeds on living coral polyps by spreading its stomach over the coral and digesting an area nearly as large as its body disc.

The starfish grows to a large size, commonly 30–40 cm across, and rarely up to 60 cm. The general colour is usually blue grey with a reddish tint when the tiny breathing vesicles are expanded, while the spines are red to orange. It has a large disc with about 15 to18 fairly short arms radiating from it. The disc and arms are covered with a soft skin and stout, hinged spines 2–3 cm long, each with a three sided blade at the tip which can penetrate wetsuits and heavy gloves. A thin skin (the epidermis) which contains glandular cells covers the spines. One type of glandular cell (acidophilic) is thought to produce the toxic substances. The venom consists of a neurotoxin, histamine-like substances and a protein toxin.

In Western Australia, crown-of-thorns starfish are common in the Dampier Archipelago, but rare on coral reefs as far south as Coral Bay on the Cape Range Peninsula. They sometimes lie in the open but are more often concealed under plate corals or ledges during the day, their colour blending with the background so they may be unwittingly touched or trodden upon.

The spines cause puncture wounds and the tips often break off and become embedded. The tissues contain toxic saponins, and are poisonous not only to humans but also to insects and soil organisms, suppressing plant growth. The starfish therefore cannot be used either for food or fertiliser.

INJURY PREVENTION

- Wear gloves and protective clothing while diving or snorkelling on coral reefs.
- Take extreme care if near crown-of-thorns starfish.
- Wear boots if walking on a tropical coral reef.
- Look first before putting a hand under a ledge.

Clay Bryce

▲ The crown-of-thorns starfish, *Acanthaster planci*, has razor sharp spines covered by thin skin containing venom glands.

SYMPTOMS

- Puncture wound from a spine is intensely painful with throbbing, and bleeds freely.
- Swelling, redness, tingling heat and numbness of area surrounding wound.
- Stinging by 10 or more spines may result in vomiting which can recur every few hours for several days.
- Dermal tissue, mucus, terrestrial or marine bacteria as well as venom will be deposited in wound.
- If infection develops in puncture wounds, lymph glands in armpit or groin may become tender or swollen.
- Often spine tips break off in wound resulting in complications, which may develop weeks or even months later. A spine tip in a finger can result in swelling and stiffness caused by growth of granulation tissue typical of a foreign body reaction. Tissues may form a hard ball around foreign body or bone destroying (osteolytic) process may cause narrowing of a joint by destruction of cartilage. Surgery is needed to correct both of these effects.

Barry Wilson

▲ The underside of the crown-of-thorns starfish, *Acanthaster planci*, showing the central mouth and the tube feet on the arms.

- Abnormal swelling and itching may indicate an allergic reaction, which may increase with subsequent stingings; increased susceptibility (anaphylaxis) is possible.
- Protracted acute illness can result from multiple punctures (Edmonds (1989).
- Intense pain can cause disorientation and panic in divers, leading to unsafe ascent.

TREATMENT

The steps below can be followed for treatment of minor wounds, and for first aid. Medical treatment should be sought for all but most minor injuries, especially if any spine fragments remain embedded. Any suspicion that a piece of spine is actually in the joint warrants immediate medical attention, and if confirmed, surgery (Williamson *et al.* 1996: 317)

- Wash wound carefully to remove traces of mucus and dermal tissue.
- Immerse affected part in water as hot as can be borne. Most marine poisons are destroyed by moderate heat — problem is to get heat to the depth where the venom is deposited without scalding. Minimise risk of scalding by testing water temperature with unaffected part of body before immersing affected part.
- Strong painkillers may be needed.
- If protracted vomiting occurs, try to stop with antiemetics, and make sure victim does not become dehydrated.
- Antihistamines should be given if there is an allergic reaction.

Refer to treating infections and allergies, page 219–223.

CASE REPORT

Puncture wounds in the hand were treated by soaking the hand in cold vinegar and water, which resulted in a decrease in pain and nausea. This was followed by cold vinegar/water and occasional aluminium acetate soaks [astringent solution] repeated over two days by which time the symptoms and signs were resolved. Four days later the stung area became itchy with scaly red dermatitis. Systemic corticosteroids [administered by a doctor] cleared the dermatitis in six days.

(Russell & Nagabhushanam 1996)

Sea urchins

Phylum Echinodermata
Class Echinoidea

Sea urchins — also known as sea eggs — are circular or oval, spherical or flattened echinoderms with a shell of closely fitting plates, covered by a thin skin (Figure 13). Spines, from the thinness of a needle to the thickness of a pencil, and from a few millimetres to more than 20 cm in length, project from the shell. The spines can turn on rounded knobs, allowing movement for defence or locomotion. Like other echinoderms, hydraulically operated tube feet with suckers at the tips are the main means of movement. The mouth on the lower surface has a complex jaw with five strong teeth, which are used to crop the algae on which many feed. As well as spines, sea urchins have minute pincer organs, or pedicellariae, which remove foreign bodies from the skin.

Most sea urchins can be handled safely but some have sharp venomous spines (see Figure 14). Others have highly venomous pedicellariae (Figure 15).

Although painful, injuries from non-venomous sea urchin spines are not harmful. If the spines cannot be entirely pulled out the small fragments remaining will be gradually absorbed. Refer to page 186 for treating minor echinoderm injuries.

FIGURE 13:

Diagrammatic section through a sea urchin showing spines on the left and tube feet on the right. In life the gut and five gonads occupy most of the body cavity. (After Buchsbaum 1951)

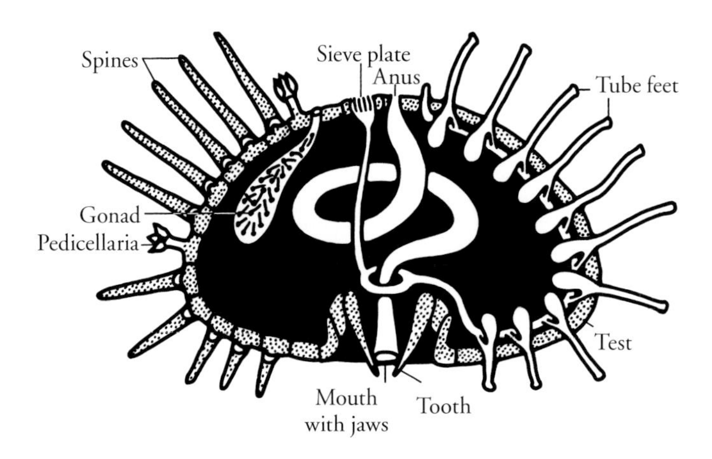

SEA URCHINS WITH VENOMOUS LONG SPINES

Species of *Diadema*, *Echinothrix* and *Centrostephanus* all belong to the family Diadematidae and are known as needle-spined urchins.

Diadema spp.

Phylum Echinodermata
Class Echinoidea
Family Diadematidae

Species of *Diadema* are tropical, needle-spined sea urchins, which are found on coral reefs, rocky coasts and on the sandy seabed in shallow water, from low tide level to a depth of about 70 metres. They have a flattened circular shell up to 10 cm wide, covered by black skin. The spines are hollow, brittle and very slender, reaching about 25 cm in length. Shorter and thinner secondary spines are scattered among them. These urchins are usually black in the adult (occasionally with some white spines) but in juveniles the spines have black and white bands.

Diadema setosum has a reddish ring around the anus in the centre of the upper surface. Five blue spots and radiating white lines contrast with the black skin. The light sensitive spots react to shadows or nearby

Clay Bryce

▲ The needle-spined urchins *Diadema* spp. can cause painful wounds with their spines, which readily penetrate the skin and break off.

movements, causing the spines to wave and converge on an intruder.

Diadema savignyi differs only by having a black anal area, a blue ring around the centre of the upper surface with five pairs of electric blue lines radiating from it, and the white spots may be absent. Both *Diadema setosum* and *D. savignyi* are common in north western Australia and as far south as the coral reefs of the Houtman Abrolhos islands off Geraldton. *Diadema savignyi* is occasionally found at Rottnest Island. Both species may group in large colonies during the day usually on the bed of sandy lagoons or under rocks, and disperse at night to feed on algae.

The long needle spines are very fragile and very sharp, and readily penetrate the skin and break off. Their rough surface and brittle texture make them almost impossible to remove (see Figure 14, page 198). Venom from the spines of a Caribbean species of *Diadema* has been isolated (see Williamson *et al.* 1996). It is composed of proteins, steroid glycosides and inflammatory substances. To date no venom has been found in the spines of Indo-Pacific species.

SYMPTOMS

- Severe pain, lasting from half an hour to 4 hours.
- Swelling and inflammation may occur.
- Fragments of spines may be visible under the skin.
- Injury may be complicated by a foreign body reaction to small, detached pieces of spine or the skin covering the spine.

TREATMENT

Medical treatment should be sought for major injuries. *Diadema* spines are extremely brittle and so a spine can rarely be removed intact. Usually spine fragments of *Diadema* are absorbed without harm, or work their way out. The steps below can be followed for treatment of minor wounds, and for first aid.

- Try to dissolve spine fragments by soaking affected part in vinegar.
- Usually spine fragments of *Diadema* are absorbed without harm, or work their way out.
- Apply antiseptics to the affected area.

Refer to treating minor echinoderm wounds, page 186.
Refer to Infections and Allergies, page 219–223.

Echinothrix spp.

Phylum Echinodermata
Class Echinoidea
Family Diadematidae

Sea urchin species of *Echinothrix* are readily distinguished from *Diadema* by five bands of very fine, golden brown to black spines extending from the upper to the lower surface of the body protecting the tube feet. The other primary spines are shorter than those of *Diadema* and in adults are less than half as long as the shell width which may be up to 15 cm. In small examples the spines are often longer than the width of the shell.

Echinothrix calamaris (below), particularly a young specimen, can be a beautiful sea urchin with black, white or green, often banded, primary spines and bands of very fine, golden brown shorter spines. Very large specimens are often entirely black. Small white plates are embedded in the skin around the anus (a good diagnostic characteristic).

Clay Bryce

▲ *Echinothrix* spp. have shorter primary spines than *Diadema* and they also have very fine, barbed venomous spines which are grouped in five bands.

FIGURE 14:

Tips of spines of (a) *Diadema setosum* (x33), (b) *Echinothrix diadema* (x33), (c) *Araeosoma thetidis*, enclosed in a venom gland (x27), (d) *Asthenosoma varium*, with venom gland (x27) (After Mortensen 1935, 1940).

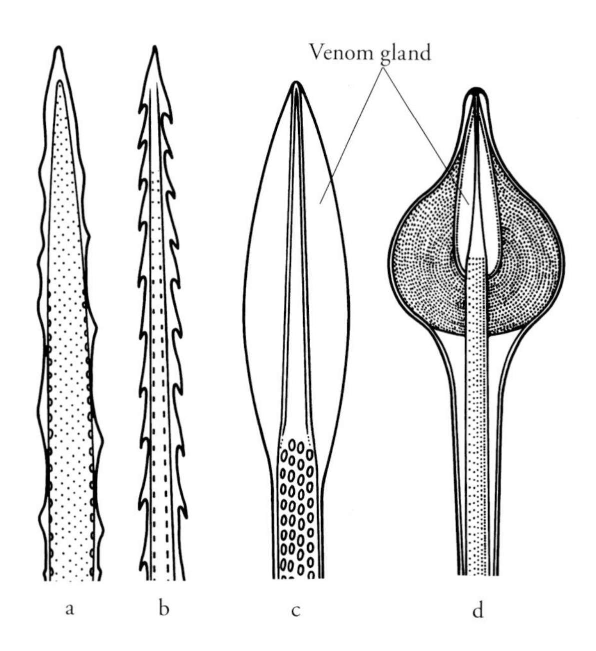

The primary spines are hollow with a large cavity, making them extremely brittle. They are covered by whorls of minute scales, the free ends of which point towards the spine tip. The fine spines have barbs pointing away from their tips (Figure 14 b) which make them impossible to pull out if they become embedded in a victim. The tips of these are encased in skin containing venom glands.

Echinothrix diadema are less brightly coloured than *E. calamaris* and small specimens often have dark green primary spines banded with black, while the spines of large specimens are usually dark coloured. The hollow primary spines have only a small cavity, making them less brittle than those of *E. calamaris*, and the scales are on ridges running along the spines and not in whorls around them. There are no white plates near the anus. The fine spines are venomous.

Both species live in shallow water on coral reefs, often partly concealed under boulders or ledges. They are both found throughout the tropical Indo-West Pacific and on the northern, eastern and western coasts of Australia southwards to the Houtman Abrolhos in Western Australia.

Venom glands at the tips of the fine spines of both species contain noradrenaline, which affects blood pressure, and a protein, which is believed to cause the intense pain.

SYMPTOMS

- Immediate severe burning pain and numbness, which becomes a throbbing pain as redness and swelling develop.
- Fragments of spines may be visible particularly as the surrounding skin can become stained blue.
- Secondary infection of the puncture wounds is common.
- Delayed reactions from embedded spine fragments are likely to occur.

TREATMENT

Since the spines are venomous, barbed and not removable, and are not absorbed, treatment differs from that of *Diadema* injuries.

- Seek medical attention without delay — prompt surgical removal is desirable, and further medical treatment may include prescription of general antibiotics directed against secondary infection, and analgesics and anaesthetics for pain relief. Prior to medical treatment the injured part should be handled as little as possible to avoid fragmentation of the spines.
- As a first aid measure, immerse affected part in water as hot as can be borne as most marine poisons are destroyed by moderate heat. Problem is getting the heat to the depth where the venom is without scalding. Minimise risk of scalding by testing temperature of water with an unaffected part of body before immersing affected part.

Refer to Infections and Allergies, page 219–223.

CASE REPORT 1

In a West Australian case, a spine in a fingertip caused intense pain and numbness which spread to the whole hand within a minute with throbbing pain running up the arm.

After half an hour the whole arm was numb, but after an hour the numbness decreased, persisting in the finger for 24 hours.

The area around the entry point of the spine turned bright blue initially, and then the colour spread to the whole finger, fading to purple and retreating to the fingertip after half an hour. Ten days later tingling and pain with some swelling of the fingertip returned but disappeared after three days.

(W.H. Butler, unpublished)

CASE REPORT 2

Earle (1941) noted that 'fragments of spines in a toe caused pain and swelling of the foot three months after penetration, and the pieces of the spine had to be removed under general anaesthetic'. However, this was erroneously attributed to *Diadema* as Earle's illustration shows a species of *Echinothrix*, and confusion between the symptoms of *Diadema* and *Echinothrix* have persisted in the literature. The spines of *Echinothrix* are not absorbed like those of *Diadema*, and unremoved fragments may be seen by x-ray many months after the original injury.

Southern needle-spined urchin

Centrostephanus tenuispinus H. L. Clark, 1914

Phylum Echinodermata
Class Echinoidea
Family Diadematidae

The southern needle spined urchin is large — up to 11 cm across — and has hollow scaly spines 7 cm long. The shell is covered with a dark skin of reddish-brown to dull green or purple, and the spines are lavender grey to nearly black.

Centrostephanus is common on the south western coast of Australia, from Shark Bay to South Australia. It is not found elsewhere, although a related species lives on the east coast of Australia. It lives among rocks or under ledges on limestone reefs below low tide level to a depth of 100 metres.

Unlike other diadematid urchins, *Centrostephanus* has pedicellariae (pincer organs) with venom glands. No case of stinging by this species has been reported but it should be handled with care because of its needle spines and venomous pedicellariae.

TREATMENT

Medical treatment should be sought for all but the most minor injuries, especially if any spine fragments remain embedded. *Centrostephanus* spines are extremely brittle and so a spine can rarely be removed intact. Usually spine fragments of *Centrostephanus* will be absorbed without harm, or will work their way out. The steps below can be followed for treatment of minor wounds, and for first aid.

- Immerse affected part in water as hot as can be borne to ease the immediate pain from a spine puncture and for lasting relief, repeat as water cools. Minimise risk of scalding by testing temperature of water with an unaffected part of body before immersing affected part.
- If not deeply embedded, spine fragments may be dissolved by soaking injured area in vinegar.

For further steps in treating minor echinoderm wounds refer to page 186 and also see treating Diadema injuries on page 196.

Clay Bryce

▲ The common needle spined urchin in south-western Australia is *Centrostephanus tenuispinus*. Although it has sharp brittle spines and venomous pedicellariae no injury from this species has been recorded.

SEA URCHINS WITH VENOMOUS SHORT SPINES

Tam O'Shanter urchins

Phylum Echinodermata

Class Echinoidea

Families Echinothuriidae and Phormosomatidae

Tam O'Shanter urchins, *Asthenosoma*, *Phormosoma* and *Araeosoma* species, are large (up to 14 cm across), brightly coloured, short spined sea urchins with a thin shell which collapses out of water making them look like berets, hence the common name.

All have highly venomous spines, which are fairly short, thin, extremely sharp and able to penetrate most types of gloves (see Figure 14 c, d). In

Clay Bryce

▲ *Asthenosoma* spp. have short, fine, very sharp spines encased in venom glands. They sting severely at the slightest touch.

Asthenosoma the spines of the upper side are covered by an inflated skin sheath, sometimes white but often pale coloured at the base with purplish bands shading from pale to brilliant electric blue at the tips. The spines each have a single large poison sac at the tip and venom is injected when the spines penetrate the skin and, usually, break off in the wound. The composition of the venom is not known.

Species of *Asthenosoma* occur from low tide level to about 150 metres. Two Indo Pacific species, *A. varium* and *A. ijimai*, are found in northern Australia, southwards to the coral reefs of the Houtman Abrolhos.

Two species of *Araeosoma*, *A. owstoni* and *A. tesselatum*, have been trawled from a depth of 150 to 400 metres on the North West Shelf. They are similar in appearance to species of *Asthenosoma*. *Phormosoma bursarium* has also been trawled below 170 metres from the North West Shelf and slope.

INJURY PREVENTION

- Do not pick up these sea urchins, even with heavy gloves.
- Beware of these urchins when sorting trawl catches although trawled animals are less harmful as most of the spines will be incomplete.

SYMPTOMS

- Immediate intense pain with swelling, lasting several hours.

TREATMENT

Medical treatment should be sought for all but the most minor injuries, especially if any spine fragments remain embedded. Surgical removal of spine fragments is desirable. The steps below can be followed for treatment of minor wounds, and for first aid.

- Immerse affected part in water as hot as can be borne to ease the immediate pain from a spine puncture and for lasting relief, repeat as water cools. Minimise risk of scalding by testing temperature of water with an unaffected part of body before immersing affected part.
- Apply local anaesthetics.

For further steps in treating minor echinoderm wounds refer to page 186.
Refer to Infections and Allergies, page 219–223.

SEA URCHINS WITH VENOMOUS PEDICELLARIAE

Phylum Echinodermata
Class Echinoidea

The pincer organs, or pedicellariae, of sea urchins are usually minute, often pinhead sized, seizing organs scattered among the spines. Some are used for defence while other types clean the body surface. They consist of three hard jaws on a flexible stalk. Pedicellariae occur in various shapes and sizes, characteristic of each species, but one kind, the globiferous pedicellariae (Figure 15), has venom glands connected to sharp, fang like teeth at the ends of the jaws.

Many sea urchins have globiferous pedicellariae for defence but they are usually too small to be harmful to humans. However one species, the flower urchin, *Toxopneustes pileolus*, has very large, highly venomous pedicellariae and is dangerous. Several other species are less toxic but should be handled with care.

FIGURE 15:

A venomous pedicellaria of the sea urchin *Toxopneustes* showing the needle sharp fangs connected to venom glands in the three jaws (after Mortensen 1943).

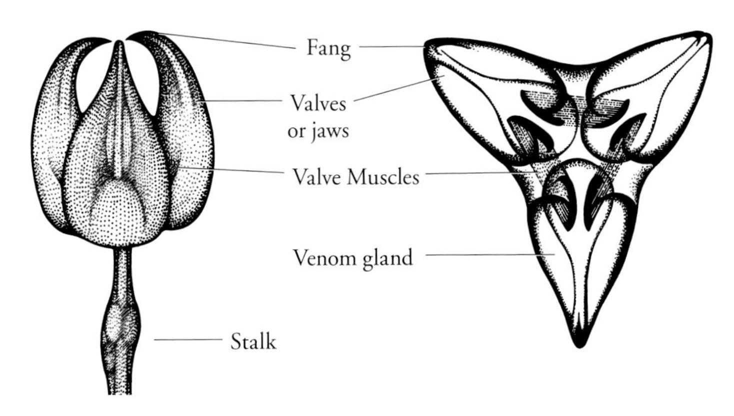

Flower urchin

Toxopneustes pileolus (Lamarck, 1816)

Phylum Echinodermata
Class Echinoidea
Family Toxopneustidae

The body of the tropical flower urchin, *Toxopneustes pileolus*, is a low round dome, with a concave underside. It can be up to 15 cm wide, but

6–10 cm is more usual. The spines of the upper side are 6–8 mm long, and those on the lower side 10–12 mm long. The spines are fairly robust, taper to a point, and are in shades of red, green or white, or sometimes bands of all three colours.

In life, the spines are almost concealed by the open, flower-like globiferous pedicellariae each of which has a rounded triangular shape up to 3 mm across (Figure 15 and below). When sensory hairs on the inner side of the jaws of the pedicellariae are stimulated, the jaws close and inject venom through the fangs. They cannot bite through the thick skin of the palm of the hand but the side of a finger or tender skin is susceptible. Bites by several pedicellariae cause severe pain and general symptoms. The venom is basic in nature and interferes with nerve muscle conduction and so paralyses the muscles. It also increases the leakage of fluid from blood capillaries, which can be reversed by antihistamine drugs.

Drowning, as a result of being bitten, has been reported in Japan. The toxin was purified by Nakagawa *et al.* (1991) who named it Contractin A. Takei *et al.* (1991) isolated three protein fractions and found that one causes histamine release (see Russell & Nagabhushanam 1996).

The flower urchin is a widespread Indo-Pacific species found in northern

Clay Bryce

▲ The short spines of the flower urchin, *Toxopneustes pileolus*, are hidden by the large, open, flower-like but highly venomous pedicellariae.

Australia and as far south as Montague Island in New South Wales and Shark Bay in Western Australia. It lives on coral reefs and sand or mud from the intertidal zone to a depth of 90 metres. (Endean 1961)

INJURY PREVENTION

- Wear gloves if you need to handle the flower urchin.

SYMPTOMS

- Pain, severe and spreading.
- Numbness, lasting up to an hour.
- Collapse, paralysis and breathing difficulty or shallow breathing lasting up to six hours.

TREATMENT

- Seek medical assistance. Medical action may include regional block anaesthesia or intravenous pethidine for pain relief, and treatment for shock.
- Pain relief is paramount. As a first aid measure, immerse affected part in water as hot as can be borne as most marine poisons are destroyed by moderate heat. Problem is getting heat to depth where venom is without scalding. Minimise risk of scalding by testing temperature of water with an unaffected part of body before immersing affected part.
- Carefully remove any pedicellariae adhering to skin.
- If breathing becomes shallow with skin turning bluish in colour, start and maintain expired air resuscitation (see pages 227–228).
- If expired air resuscitation does not correct patient's colour or if no pulse or heartbeat can be felt, also apply cardiopulmonary resuscitation (see pages 229–231).
- Antihistamine drugs may reverse some of the effects.

CASE REPORT

A bite caused by seven or eight pedicellariae on the side of a finger caused severe pain followed by giddiness, difficult breathing, paralysis of the lips, tongue and eyelids, and relaxation of the muscles in the limbs.

After 15 minutes the pain gradually diminished, and disappeared after an hour. The face was paralysed for about six hours.

(Fujiwara 1935, quoted by Mortensen 1943)

Tripneustes gratilla (Linnaeus, 1758)

Phylum Echinodermata
Class Echinoidea
Family Toxopneustidae

This very common tropical Indo West Pacific sea urchin, sometimes known as the cake urchin, lives on coral reefs and limestone platforms as far south as Rottnest Island in Western Australia. It has a depth range from the intertidal zone to about 70 metres. *Tripneustes* is usually concealed under ledges, or covers itself with fragments of algae and shells.

The beach-washed shell is often pale lavender in colour but when alive, short white spines stand out conspicuously against the dark skin covering the shell. In other parts of the Indo Pacific the spines may be orange or even black. Amongst the spines are many very small globiferous pedicellariae but with normal handling these are too small to bite. However, if one becomes attached to the tender skin of a finger or inside of the arm it may sting quite severely.

The venom, analysed by Feigen *et al.* (1966), is acidic and contains a water-soluble protein which lowers blood pressure and ruptures red blood cells. It also contains a chemical which affects the heart, and acts as an enzyme, causing histamine release. The venom is deactivated by heating

Loisette Marsh

▲ *Tripneustes gratilla*, which also has venomous pedicellariae, is common on coral reefs. However this specimen, flanked by colonies of *Palythoa*, was photographed at Rottnest Island, near Fremantle.

to 45–47.5°C. The gonads of *Tripneustes gratilla* are edible, but should be well washed to remove any pedicellariae that may adhere to them.

INJURY PREVENTION

- Wear gloves when handling
- Wash the gonads thoroughly before eating.

SYMPTOMS

- Immediate pain, localised swelling and redness, and an aching sensation in the affected part.
- Allergic reactions may follow subsequent stingings.

TREATMENT

- Immerse affected part in water as hot as can be borne as *Tripneustes* toxin is destroyed by moderate heat. Problem is getting heat to depth where venom is without scalding. Minimise risk of scalding by testing temperature of water with an unaffected part of body before immersing affected part.
- Carefully remove any pedicellariae adhering to skin.
- Treat allergic reactions with antihistamine.
- Seek medical assistance if pain is excessive or allergic reactions are severe.

Refer to Infections and Allergies, pages 219–223.

CASE REPORT 1

A sting on the tender skin inside the elbow by pedicellariae on a fragment of shell was more painful than nettles and lasted several hours. The resulting wound, similar to a burn, did not heal for more than a month.

(Mortensen 1943)

CASE REPORT 2

The pain from a sting by a single pedicellaria was similar to a bee sting. A minute swelling at the site of each of the three puncture marks was followed by a weal 1 cm across. The symptoms lasted about one hour and there were no general effects. Subsequent stingings reacted more severely, lasting up to eight hours, with a weal 12 cm across indicating increasing histamine response.

(Alender & Russell 1966)

Loisette Marsh

▲ A slate pencil sea urchin, *Heterocentrotus mammillatus*, on a coral reef.

Slate pencil urchin

Heterocentrotus mammillatus (Linnaeus, 1758)

Phylum Echinodermata
Class Echinoidea
Family Echinometridae

The slate pencil sea urchin, *Heterocentrotus mammillatus*, found on coral reefs, is only known at present in Western Australia from atolls far off the north western coast, but it is moderately common on the Great Barrier Reef off Queensland. Stings from its venomous globiferous pedicellariae have been reported.

Salmacis sphaeroides (Linnaeus, 1758)

Phylum Echinodermata
Class Echinoidea
Family Temnopleuridae

This short-spined sea urchin is fairly common on shallower parts of the North West Shelf off Western Australia, and is often taken in fish and prawn trawls. It has fairly large globiferous pedicellariae, but stings have been reported from the spines.

Barry Wilson

▲ *Salmacis sphaeroides* is often trawled on the North West Shelf off Western Australia. This animal was photographed emitting clouds of sperm.

SYMPTOMS FOR SLATE PENCIL URCHIN AND *SALMACIS SPHAEROIDES*

- Immediate pain and an aching sensation in affected part.
- Localised swelling and redness.
- Allergic reactions may follow subsequent stingings.

TREATMENT FOR SLATE PENCIL URCHIN AND *SALMACIS SPHAEROIDES*

- Immerse affected part in water as hot as can be borne as most marine poisons are destroyed by moderate heat. Problem is getting heat to depth where venom is without scalding. Minimise risk of scalding by testing temperature of water with an unaffected part of body before immersing affected part.
- Carefully remove any pedicellariae adhering to skin.
- Treat allergic reactions with antihistamine.
- Seek medical assistance if pain is excessive or allergic reactions are severe.

Refer to Infections and Allergies, pages 219–223.

Sea cucumbers

Phylum Echinodermata
Class Holothuroidea

Holothurians, or sea cucumbers, are sausage shaped animals, from a few centimetres to over a metre in length (see below). Some lie still on the sandy seabed using short sticky tentacles to collect food particles from the sand, while others spread long branching tentacles in the water to trap plankton. Like other echinoderms they move about on tube feet which extend through the body wall.

Sea cucumbers are most abundant in tropical seas where a number of shallow water species are collected and processed for food, used by Chinese, Japanese and Pacific Islanders. The large species used for food have thick body walls and belong to the families Holothuriidae and Stichopodidae. The prepared product is known as trepang and bêche-de-mer.

Clay Bryce

▲ *Actinopyga mauritiana* is one of many large sea cucumbers known as bêche-de-mer, and as trepang when prepared for eating. All the edible species are toxic unless correctly prepared.

Over 40 species of sea cucumber have been tested, including all those used for food, and found to contain toxic saponins named Holothurin A and B, which irreversibly block nervous and muscular action and cause the rupture of red blood cells. The poisons dissolve in water but are not destroyed by heat. Dried sea cucumber products have little toxicity (Hashimoto 1977) as lengthy contact with water during processing leaches most of the poison from the body wall. Processing includes gutting, boiling for up to 1.5 hours, washing and drying.

When disturbed, many species of sea cucumber rapidly eject sticky white threads, or Cuvierian tubules, from the anus. These are highly toxic and in some species, for example those of the large leopard (or tiger) fish, *Bohadschia argus*, cause inflammation if they touch the skin.

Holothurians are most common on the tropical reefs of Western Australia but some species occur along the western and southern coasts.

Clay Bryce

▲ Sticky Cuvierian tubules are ejected from the anus of *Bohadschia argus*.

INJURY PREVENTION

- Do not eat bêche-de-mer unless they have been correctly prepared.
- Avoid contact with the white threads ejected from the anus. If these are touched, be careful not to then touch face, especially eyes, as toxin is highly irritant.

SYMPTOMS

- Severe illness or even death can result from eating raw sea cucumbers.
- The white tubules can cause contact dermatitis.

TREATMENT

- Seek immediate medical attention if raw sea cucumber has been eaten.
- Seek medical attention if eyes are affected, with referral to an eye specialist.
- Corticosteroid cream may help with dermatitis.

Animal studies suggest that anticholinesterase therapy may be effective for poisoning from eating sea cucumbers.

(Friess 1963)

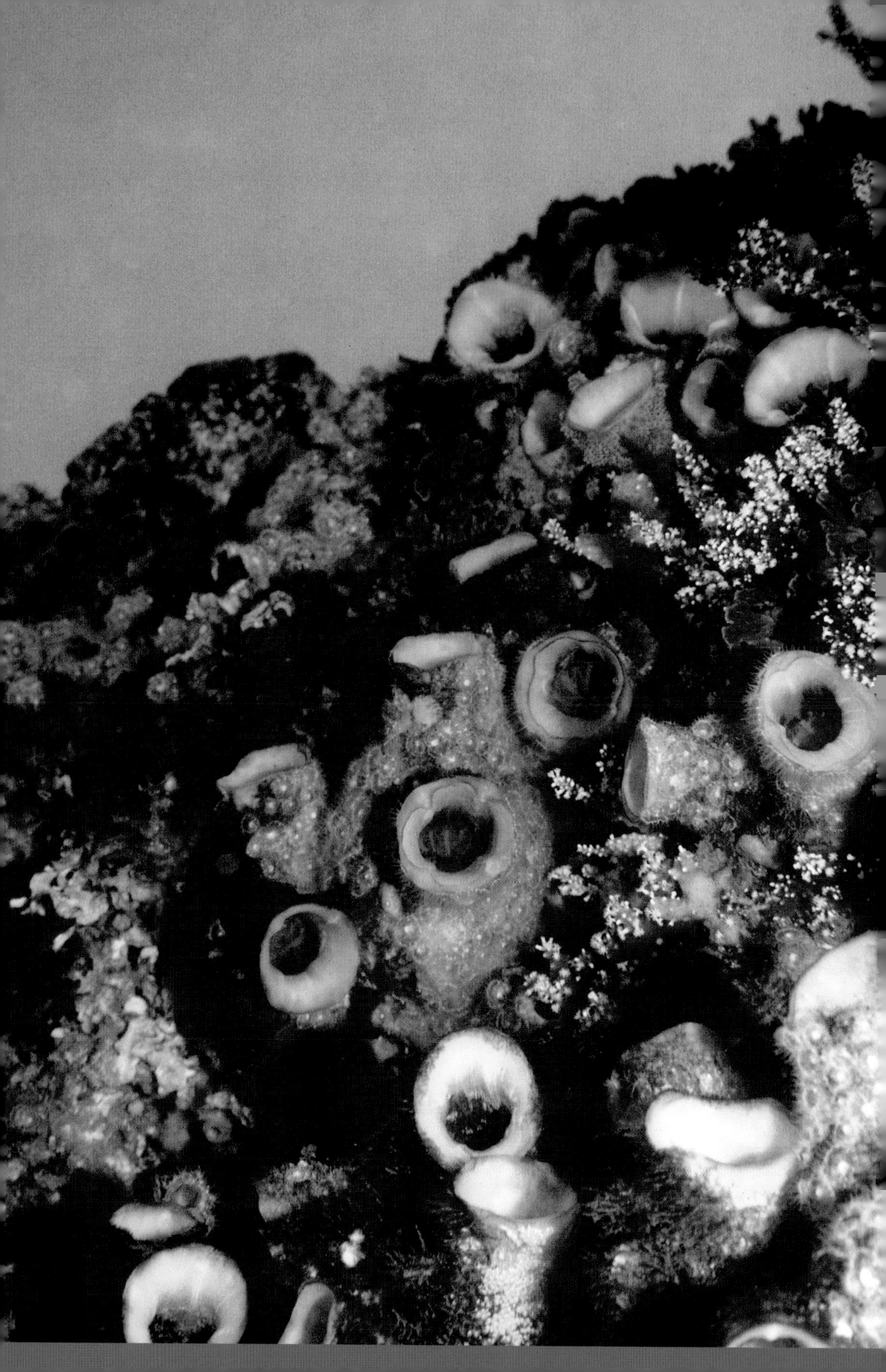

Sea Squirts

Phylum Tunicata
Class Ascidiacea
Family Pyuridae

Sea squirts, or ascidians, are quite complex animals, deceptively simple in appearance. They are sac like, solitary or colonial animals, which live attached to rocks or jetty piles and are some times used for fishing bait.

The common name arises from their ability to contract the muscular inner body wall squirting a jet of water through one or both of the external openings. Water is taken in through an inhalant siphon to a basket-like pharynx where minute food particles are sieved out and passed on to the stomach. Oxygen is extracted as water passes through the pharynx into a surrounding chamber, which leads to an exhalent siphon from which it is expelled (see Figure 16).

The soft internal organs are enclosed in a thin but muscular body wall, protected by an outer 'tunic' composed of a cellulose-like substance, which in some species is thick and tough.

In the large pink sea squirt, *Herdmania grandis* (see page 217), which may be 10–15 cm high and as much across, the inner body wall is strengthened with a mat of calcareous needle like spicules which can painfully penetrate the hands if the animal is cut or damaged. This is a southern Australian species which attaches to rocks in shallow water.

Another species, *Pyura stolonifera*, known as cunjevoi, is used as bait and sometimes food on the east coast of Australia and although found in

Clay Bryce

◀ A reef wall at Albany with *Herdmania grandis*.

FIGURE 16:

A sea squirt with the tunic and body wall removed from one side to show the internal organs. Water passes through the holes in the pharynx to the atrium and out by the exhalent siphon while food is collected on one side of the pharynx and passed down to the stomach. The anus and reproductive ducts all open into the atrium.

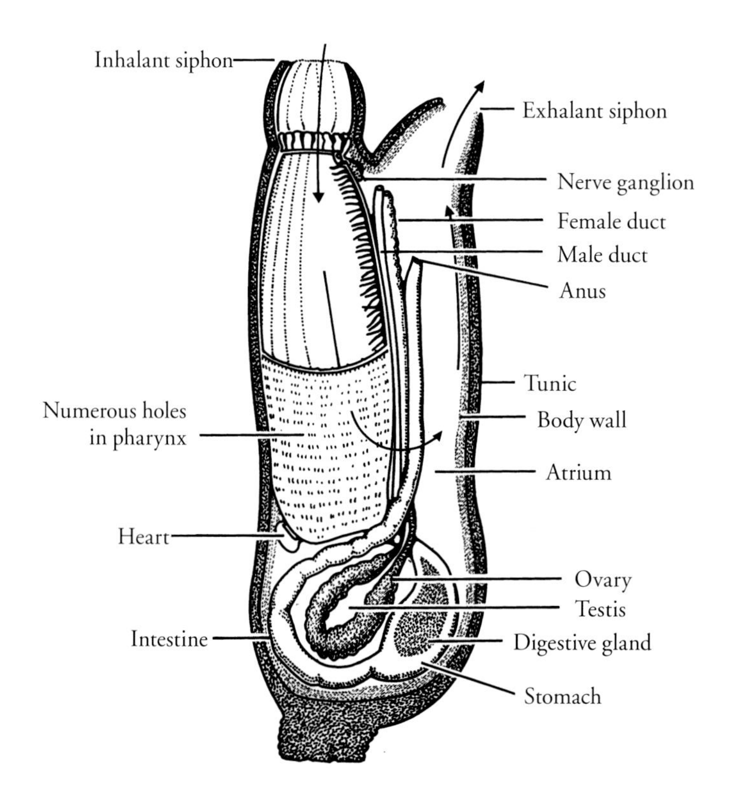

Western Australia, it is uncommon. *H. grandis* looks equally inviting to eat but the numerous spicules can painfully penetrate the mouth if one attempts to eat it.

Ascidians lack a backbone but have tadpole like larvae with an elastic skeletal rod (notochord) and gill clefts similar to those possessed by larval vertebrate animals. For these reasons they were included in the phylum Chordata to which fishes and humans belong, but are now regarded as a separate phylum, Tunicata.

INJURY PREVENTION

- Wear gloves if handling sea squirts.

SYMPTOMS

- Pain and irritation from penetration of spicules (but no symptoms associated with poisoning because no venoms or nematocysts).

TREATMENT

- Soak the affected part in vinegar or any weak acid, e.g. lemon juice, for a few minutes to dissolve spicules and relieve the pain. (The spicules are composed of calcium carbonate and dissolve rapidly in weak acid.)
- Apply adhesive tape to the affected area and pull off gently with the attached spicules.
- Pull spicules out with tweezers with the help of a magnifying glass.
- If inflammation follows, seek medical attention as recommended in Infections and Allergies, page 219–223.

Barry Wilson

▲ Many species of ascidians live among corals on tropical reefs but the large pink sea squirt, *Herdmania grandis* is a southern Australian species. It has the inner body wall strengthened with fine calcareous spicules which can penetrate the skin of the hands.

Infections and Allergies
Bacterial Infections of Marine Injuries

Secondary infection can cause maritime injuries to heal slowly. Low grade infection can result in exposure to the bacillus

Mycobacterium marinum which can cause extensive lumpy swelling in and under the skin, and to the bacterium *Erysipelothrix* which causes fish handlers' disease.

Mycobacterium marinum

Mycobacterium marinum may enter the body through lesions in the skin when exposed to water in a swimming pool, aquarium, fish tank, ship's hull or similar.

This bacillus is related to the organism which causes tuberculosis, and, like it, is very slow growing to produce a granuloma (large lumpy mass) over about two years.

The easiest type of granuloma to diagnose is a large, lumpy, non-healing ulcer. More difficult are lumps in limbs and around joints, often occurring a long time after the initial introduction of the bacillus. They have been confused with forms of arthritis or of ordinary bacterial infection.

A cyst may form, which can be aspirated (needled) or removed by surgery. The cyst may also spread to those lymph nodes related to the lesion — the armpits for the upper limbs, the groin for the lower.

Mycobacterium marinum can also cause a more diffuse infection in and under the skin.

Clay Bryce

◀ An unprotected snorkeller is at risk from coral cuts which may lead to severe infections.

TREATMENT

- Surgical removal of the mass may be necessary but should be done with caution as there is the danger that the infection could spread.
- The response to anti-tuberculosis drugs is, unfortunately, disappointing. Some antibacterials may work but courses typically need to be long. Therapy should be prescribed in consultation with a microbiologist.

FISH HANDLERS' DISEASE

Coral scratches may lead to a bacterial infection known as coral poisoning, fish handlers' disease or crayfish poisoning. The causative bacterium, *Erysipelothrix insidiosa*, is found worldwide. It enters the body through breaks in the skin and causes discomfort and sometimes illness among processors of fish, whale and meat. Skin punctures from rock lobster and fish spines are particularly prone to infection.

INJURY PREVENTION

- Wear tough puncture-resistant gloves when handling rock lobsters or fishes.
- Wear strong footwear when wading or fishing.

SYMPTOMS — MILD INFECTION

- After injury to the skin, there may be little reaction for one to seven days, during which the injury apparently heals.
- A purplish red circular area then develops around the puncture, spreading outwards at about 1 cm per day. Itching, burning or pain is felt, followed by swelling. If the area is protected against injury, but otherwise untreated, it will usually heal within three weeks. However if it is further damaged, secondary infection may result in the development of a boil or carbuncle.

SYMPTOMS — SEVERE COMPLICATIONS

- Severe complications include fever.
- Joints close to injury can stiffen.
- Septicæmia (blood poisoning).
- Septicæmia may progress to endocarditis (inflammation of membrane over valves of heart) which is obviously a serious development.

TREATMENT

- Treat all small marine cuts, particularly puncture wounds, seriously — especially for people working in relevant occupations.
- Apply antiseptic solution immediately.
- Keep the area clean and dry.
- Avoid further damage to the area.
- Medical action may include removal of fragments of coral, spines, etc, application of local antibiotic powder or ointment four times a day and a long course of penicillin or equivalent antibiotic.

Allergies

The term 'allergy' is often used in discussions on marine illnesses, and it is necessary to explain what we mean by this word. An allergic reaction is one in which a person's body responds in an unusual and usually excessive way to a foreign substance, usually a protein, which causes little or no harm to most other people. The substance may be from plants and plant foods; animals and animal foods including fishes and shellfish; animal stings whether land based or marine; and medication, topical and general.

Allergic reactions are unpredictable and may develop for a whole group of related foods or for only one type. For instance, a number of people learn from bitter experience that they are allergic to one or more types of molluscan foods, such as oysters, scallops, mussels, squid, etc. These unlucky people are affected after eating or even just touching molluscan foods which others can eat without any problem occurring.

Allergic reactions may also change with time, sometimes becoming more severe but sometimes improving.

ALLERGY PREVENTION

- Types of plants, foods and medications which have caused problems in the past should be avoided — there is a tendency for the same type of reaction to occur.
- If a person has developed an allergy to one type of plant, food or medication, then other types should be tried with great caution,

especially if they cause any swelling inside the mouth, endangering the airway.

SYMPTOMS

Some of these symptoms can result from poisoning or shock instead of from an allergy.

- Hay fever, with irritation of nose and eyes leading to sneezing and tears.
- Urticaria, when large areas of skin become red, swollen and itchy, sometimes with lumps or hives developing. A variant of this condition in which a part of the body such as a hand or part of the face swells is called angio-neurotic oedema. If severe swelling occurs inside the mouth and throat there can be a risk of suffocation.
- The most severe reaction is called anaphylactic shock, when a person goes into a state of collapse with sweating, difficulty in breathing, severe disturbance to circulation involving falling blood pressure, and fading consciousness. This dangerous condition requires urgent medical treatment as life is in jeopardy.

ANTI-ALLERGY TREATMENTS

- A characteristic of an allergic reaction is that it involves the release within the body of an inflammatory chemical called a histamine. The drugs known as antihistamines inhibit this release and can be very effective in treating allergic reactions.
- Urticaria is treated by the administration of an antihistamine by injection or by mouth for as long as the allergic reaction continues. This treatment shows its effectiveness by halting the red swelling of the skin and suppressing its intense itching. People often have antihistamine preparations which they use for other reasons such as treating hay fever, bee stings or in preventing sea sickness. These preparations can be used if medical help is not immediately available.
- Rapid relief from itching can follow the cautious subcutaneous injection of adrenalin by a doctor, which acts within seconds.
- Adrenalin is invaluable in treating life threatening anaphylactic shock.
- Cortisone, like adrenalin, is a natural body chemical and can be used,

following medical decisions and prescriptions, in several ways. Creams and ointments of hydrocortisone and related chemicals can be used to dampen ordinary and also allergic inflammation. However they should not be used if the inflammation is due to an infection. These chemicals suppress the body's natural response to infection and so their use would ultimately lead to more tissue damage. For the same reason their use by mouth or injection should be prescribed with great caution. In the treatment of shock, however, cortisone may be injected intravenously in large dosages.

First aid procedures

Recovery position
Expired air resuscitation (EAR)
Cardiopulmonary resuscitation (CPR)

RECOVERY POSITION

To place a person in recovery position

Adult and child

1. Kneel beside casualty.
2. Place farther arm at right angles to the body.
3. Please nearer arm across chest.

4. Lift nearer leg at knee so it is fully bent upwards.
5. Roll casualty away from you onto side: ensure head and neck are well supported.

6. Keep leg at right angles with knee touching the ground to prevent casualty rolling over on face.

Airway

It is essential for the casualty's airway to be clear and open so that breathing is possible. With an unconscious casualty a clear and open airway needs attention before any other injury. A blockage of the airway may be caused by:

- the tongue;
- solid or semi-solid material such food, vomit, blood or a foreign body;
- swelling or injury of the airway.

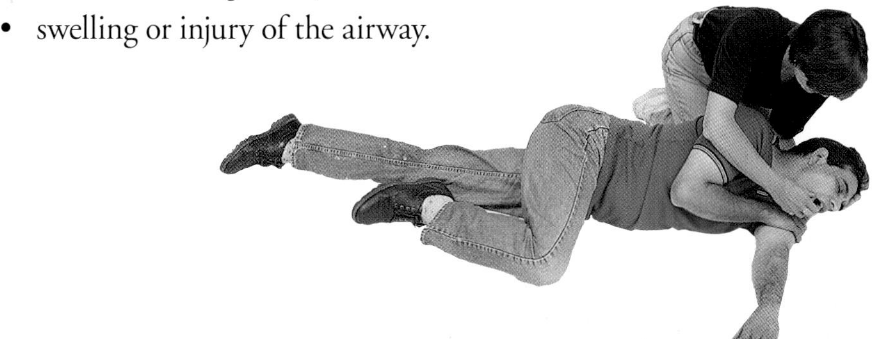

How to clear and open the airway

1. Open mouth and clear any foreign object with your fingers:
   - Only remove dentures if they are loose or broken.
2. Tilt head back gently and slightly down.
3. Place your hand high on the casualty's forehead.
4. Support chin with your other hand.
5. Gently tilt head backwards.
6. Lift jaw forward and open casualty's mouth slightly.

Breathing

To check for breathing, look, listen and feel for up to 10 seconds.

How to check for breathing

1. Look for chest movements.
2. Listen at the mouth for sounds of breathing.
3. Feel for air on your cheek.

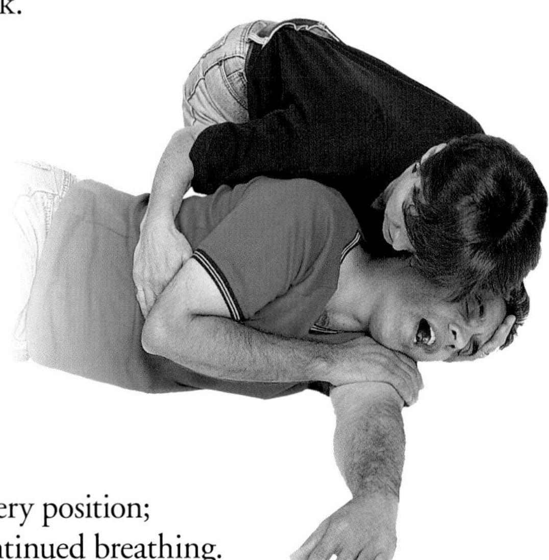

4. **If breathing:**
   - leave casualty in recovery position;
   - check regularly for continued breathing.

   OR

   If not breathing:
   - send for help – use a bystander;
   - turn casualty onto back;
   - start Expired Air Resuscitation (EAR).

EXPIRED AIR RESUSCITATION (EAR)
(Mouth-to-mouth Resuscitation)

Expired air resuscitation is used to breathe the air into the casualty to provide the oxygen needed for survival. The air you breathe out of your lungs contains about 16% oxygen. This is more than enough to keep someone alive.

Note: Expired Air Resuscitation ALONE is given to those who are not breathing but still have signs of circulation.

Adult and child from 9 years

1. Turn casualty onto back.
2. Ensure head is tilted and chin lifted.
3. Open airway:
   - pinch soft part of nose closed with index inger and thumb of hand, resting hand on forehead;
   - open casualty's mouth and maintain chin lift.

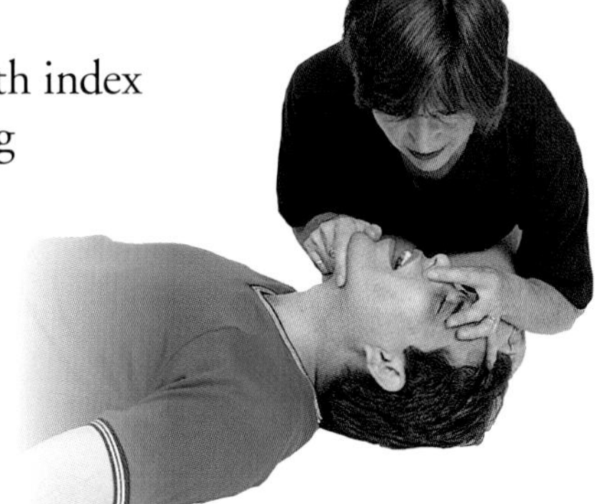

4. Take a breath and place your lips on casualty's mouth, ensuring a good seal.
5. Blow steadily into casualty's mouth for about 1.5 - 2 seconds:
   - watch for chest to rise – sign of an effective breath.

6. Maintain head tilt and chin lift.
7. Turn your mouth away from casualty and watch for chest to fall.
8. Take another breath and repeat the sequence to give at least 2 effective breaths (chest rises and falls).

   Note: *If you have trouble achieving an effective breath, recheck:*
   - mouth – remove any obstruction;
   - adequate head tilt and chin lift.

   Make up to 5 attempts in all to achieve 2 effective breaths.
   Even if unsuccessful.
9. Check for signs of circulation.

If circulation is present:

10. Continue EAR at 1 breath every 4 seconds (15 breaths per minute).
11. Check breathing and circulation about every minute.

Child 1–8 years

1. Title head back slightly and lift chin.
2. Seal child's nostrils and then open mouth.
3. Give 1 gentle breath every 3 seconds to inflate the lungs (20 breaths per minute).
4. Check breathing and circulation about every minute.

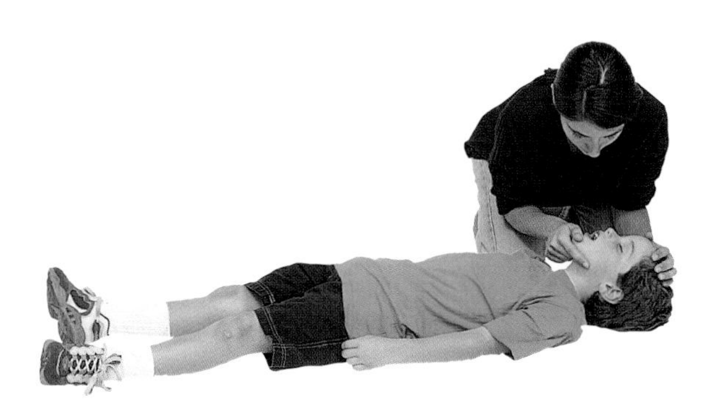

CARDIOPULMONARY RESUSCITATION (CPR)

Adult and child from 9 years

1. Kneel beside casualty, one knee level with head, the other with casualty's chest.
2. Locate lower half of breastbone (sternum):
   - find groove at the neck between collarbones;
   - find lower end of sternum by running a finger along the last rib to centre of body;
   - extend thumbs equal distances to meet in middle of sternum;

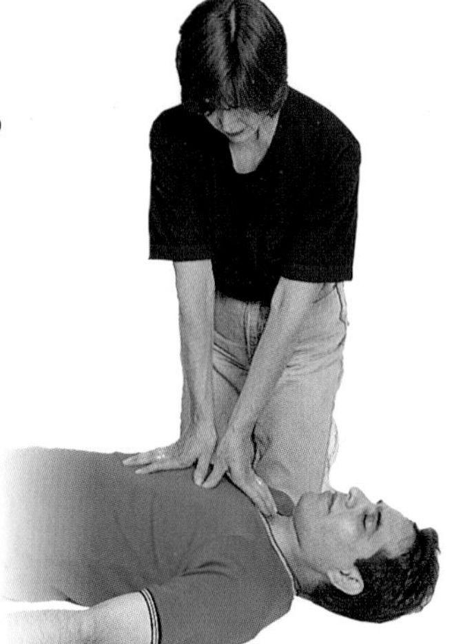

   - keep thumb of one hand in position and place heel of the other hand below it on the lower half of the sternum.

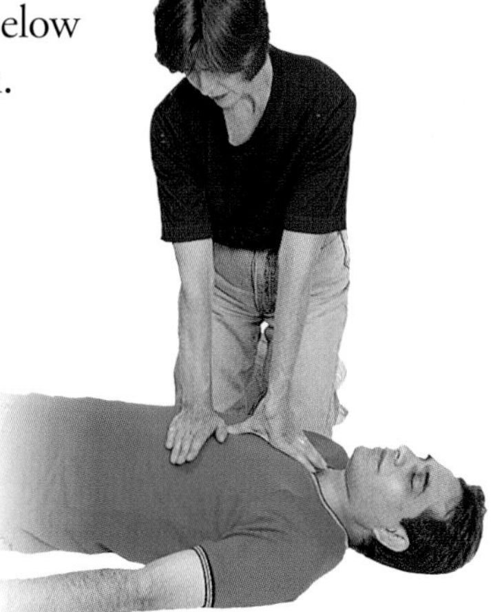

3. Place heel of the other hand on top of first.
4. Interlock fingers of both hands and raise fingers.
   - Ensure that the pressure is not applied over casualty's ribs, upper abdomen or bottom part of sternum.

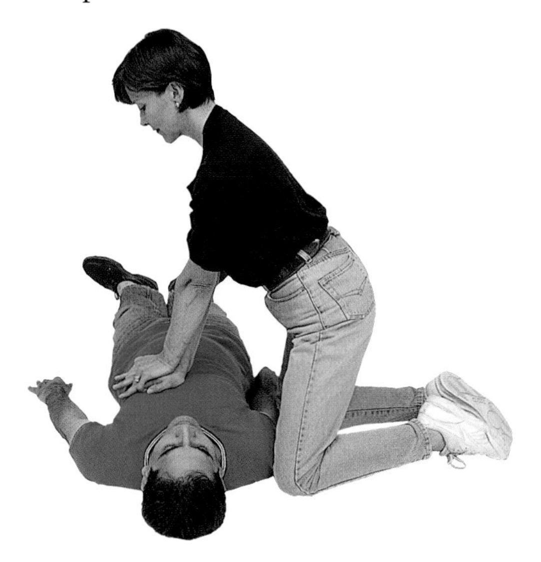

5. Position yourself vertically above the casualty's chest.
6. With your arms straight, press down on the sternum to depress it about 5 cms

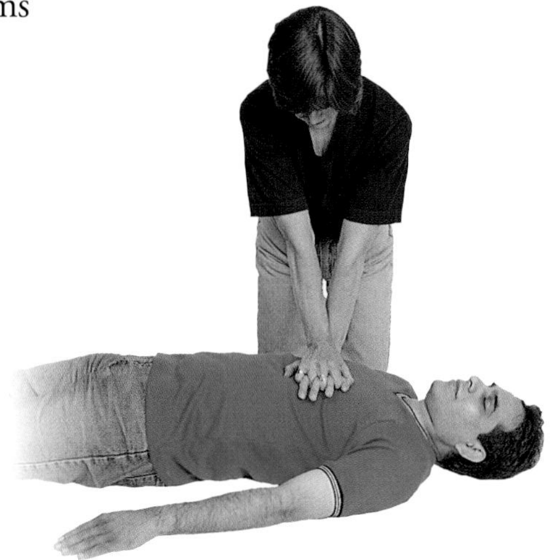

7. Release the pressure.
8. Repeat compressions at a rate of 80–100 a minute.

   Note: *Compression and release should take equal amounts of time.*

9. After 15 chest compressions, tilt head and lift chin.
10. Give 2 effective breaths.

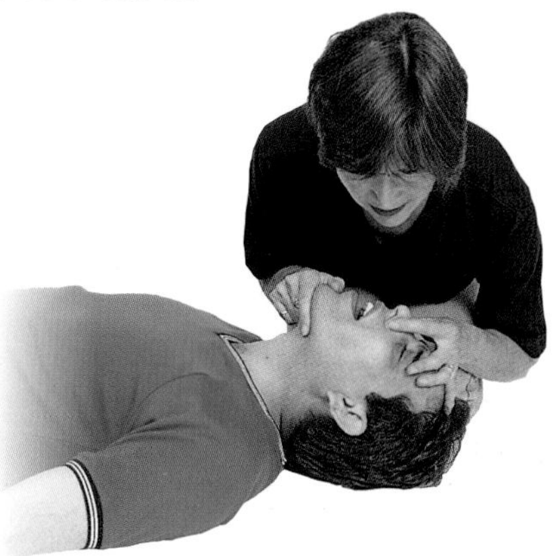

11. Return your hands immediately to correct position on sternum.
12. Continue compressions and breaths in a ratio of 15:2 at a rate of 4 cycles a minute.
13. Check for signs of circulation and breathing about every minute.

Child 1–8 years

1. Use heel of one hand over lower half of sternum to give chest compressions.
2. Compress chest approx. one third depth of chest.

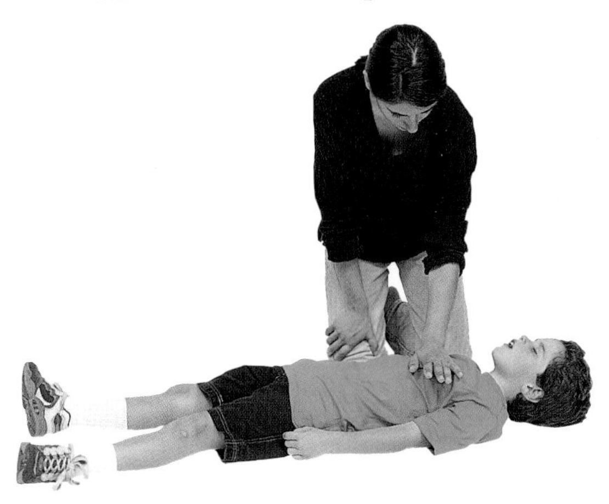

3. Give 5 chest compressions in 3 seconds followed by one breath in 2 seconds (5:1).

   Note: *Check for signs of circulation and breathing about every minute.*

Glossary

Acetylcholine: a chemical that acts as a short-acting transmitter of electric impulses from nerve to muscle at nerve-muscle junctions (synapses).

Acidophilic: ability to react with acidic dyes.

Adrenalin: A secretion of the adrenal gland which stimultates the sympathetic nervous system as raising blood pressure.

Aggregations: a group of organisms of the same or different species living closely together.

Algae: simple, mainly aquatic plants consisting of either a single cell or of many cells, possessing chlorophyll and, perhaps, other pigments but lacking internal transport systems or complex reproductive organs, e.g. seaweeds.

Allergen: a substance which can produce an allergic reaction in susceptible persons, especially after repeat exposure or contact.

Allergic reaction: an unfavourable physiological response to an allergen.

Allergy: sensitivity by some people to substances in the environment which do not affect the majority.

Amnesic: in a state of impaired memory (or amnesia).

Anaphylaxis: a major allergic reaction, usually rapidly developing, often life threatening, e.g. anaphylactic shock.

Anemone: a coelenterate which may have the cylindrical part of its body embedded in a soft substrate or adherent to solid substrates like rocks, shells, leafy algae, etc.

Antibody: a substance which the body produces in response to exposure to an antigen; it is part of the body's immune defence against the invasion of harmful foreign substances.

Anti-emetic: drugs acting to prevent nausea and vomiting – many are also antihistamines.

Anticoagulent : A substance which prevents coagulation of the blood.

Antigen: a substance not found in the body and usually a protein, for example a venom, which triggers an immune response when it enters the body; this immune response by the body includes the production of antibodies.

Antihistamines: drugs which act to counter allergic reactions.

Antiserum: a serum made from the blood of an animal or human which contains antibodies which have been developed against a specific toxin or disease.

Antivenom: a mixture of antibodies produced in the blood of an animal after it has been deliberately injected with non-harmful doses of a venom, e.g. *Chironex* venom; this antibody mixture is collected from the animal's blood, purified and concentrated, and can then be administered as an anti-venom to a human seriously affected by the relevant venom, e.g. from a Chironex sting.

Anus: the terminal opening of the digestive tract through which the faeces are expelled from the body.

Aperture: an opening.

Arm: In sea stars, crinoids (feather stars) and ophiuroids (brittle stars) a lateral projection from the central body or disc that carries a radial branch of the water vascular system and the nervous system. Also called a ray.

Asexual: without the union of gametes (as in 'asexual reproduction').

Assay: a chemical test to determine the amount or proportion of a substance in a sample.

Bacteria: very simple microscopic organisms lacking chlorophyll, many of which are parasitic.

Basophilic: able to react with basic dyes.

Bell: the 'body' of a cubozoan jellyfish.

Body whorl: the last formed and generally the largest whorl of a coiled gastropod shell.

Calcareous: chalky; consisting of calcium carbonate.

Canal: a tube or elongate excavation along which fluids may pass.

Carapace: the shelly covering of the anterior section or part of a crustacean's body.

Cardiopulmonary resuscitation (CPR): emergency life-support procedure using a combination of expired air resuscitation and external cardiac massage.

Carnivore: an animal which feeds upon other animals.

Catecholamines: Hormones and neurotransmitters released by the body under any stressful reaction or after envenomation (e.g. Irukandji) that affects the circulatory sytem, often increasing heart rate and blood pressure.

Central nervous system: the mass of nervous tissue which coordinates the activities of an animal; in vertebrate animals this consists of the brain and spinal cord.

Chitin: a strong horny substance which is resistant to chemical action and forms part of the outer covering of insects, etc.

Cilia: microscopic hairs covering whole or part of a minute organism or cell, lining the digestive or other systems of larger organisms, etc.; the concerted beating of the cilia causes movement of an organism through water or the movement of water or other substances in contact with the cilia.

Class: a unit of classification below that of a phylum; a grouping of organisms more closely related to one another than to those in other classes within a phylum.

Cobbler: a eel-tailed catfish of the family Plotosidae, possessing venomous spines on some of its fins.

Colony: a group of animals which have 'budded off' a single parent but which have incompletely separated from one another.

Contact dermatitis: skin rash (sometimes delayed) resulting from contact with an external irritant or, in susceptible people, caused by an internal reaction to antigens.

Continental shelf/slope: a gently sloping subtidal shelf around a large land mass which descends more abruptly from a depth of about 200 metres to form a continental slope.

Creepers: a common name often given to long slender gastropods of the families Cerithiidae and Potamididae which generally inhabit soft substrates and feed on detrital material from the sediment surface.

Cyanobacteria: organisms (formerly called blue-green algae) which contain various pigments giving them a green, yellow, blue, black or red colour; they may form filaments and are widely distributed in the sea, freshwater, hot springs, or cold arctic water; some cyanobacteria are symbiotic with sponges.

Dermonecrotic: causing the death of skin (also dermatonecrotic).

Dermal Tissue: The layer of skin below the epidermis.

Detritus: the decaying remains of dead animals or plants.

Diaphragm: the muscular sheet which separates the chest cavity from the abdominal cavity of mammals.

Diarrhoeal: losing fluid from the bowel.

Digestive gland: a gland which produces digestive enzymes.

Dinoflagellates: single-celled organisms which contain chlorophyll and move through fluid by the action of two whip-like flagellae (appendages).

Disc: The round or pentagonal central body of seas stars and brittle stars.

Endocarditis: inflammation of the smooth membranes lining the heart, especially those over the heart valves; bacterial endocarditis causes ulcers on these valves.

Ephyrae: tiny juvenile jellyfish which have been budded off from a polyp.

Egg capsules: the horny or calcareous coverings of one egg or a group of eggs; they may be stuck onto a hard substrate or stuck together to make an egg mass such as that of a baler shell.

Embayment: an indented sheltered coastal area.

Envenomation: the action of introducing venom into the body of another animal.

Enzyme: a protein which, in minute amounts, acts as a catalyst, promoting a chemical change in other substances without itself being used up, e.g. enzymes causing digestion of food in gut.

Exoskeleton: the hard jointed outer covering of the body of an arthropod or similar animal; muscles attached to the exoskeleton's inner surfaces cause movement of the limbs and body segments.

Expired air resuscitation (EAR): emergency life-support procedure to supply oxygen to person who is experiencing respiratory (breathing) difficulty or failure.

Exumbrellar surface: the upper surface of the bell or umbrella of a jellyfish.

Fertilisation: the union of an ovum and a sperm to form an egg.

Filter feeding: the gathering of food by filtering or sieving plankton or detritus from the water column.

Foot: the muscular locomotory organ of a mollusc such as a snail or cockle with which it moves over a surface or digs into a soft substrate.

Free-living: not parasitic.

Gastric cirri: slender filaments grouped in the corners of the stomachs of cubomedusae; on the surfaces of these cirri are nematocysts and gland cells producing digestive enzymes

Gelatinous: of a jellylike consistency.

Genus: a group of species (or, less commonly, a single species) distinguishable by their similarity to but by a degree of separation from other such species groups; the ancestors of species within a genus are thought to have been more closely related than the present species.

Gill: the respiratory organ of aquatic animals; through its surface oxygen from the water can pass into the blood and carbon dioxide can pass out into the water.

Glandular: wholly or partly consisting of glands, which are organs secreting a fluid onto the surface or into an inner cavity of an organism.

Globiferous pedicellaria: a stalked pincer organ on some sea urchins which has a globe- shaped head of 3 sharply pointed valves connected to venom glands.

Gonad: a reproductive organ that produces gametes (eggs or sperm)

Gonyautoxin: the toxic chemical or chemicals made by the single-celled marine planktonic organisms of the genus *Gonyaulax*.

Granuloma: a large lumpy mass.

Haemolytic: provoking the escape of haemoglobin from red blood corpuscles by damaging their surface membranes.

Hapalotoxin: a toxin produced by octopus species of the genus *Hapalochlaena*.

Herbivore: an animal which feeds only upon plants.

Histamine: a chemical compound produced by the body as part of allergic inflammatory reactions (v.s. antihistamine, see above).

Host: an organism which harbours a parasite; a primary host is one which supports the sexually reproductive stage of a parasite, a secondary host is one which supports other stages of a parasite's life cycle.

Ink gland or ink sac: the glandular sac of many cephalopods which secretes the dark fluid used for deflecting the attention of a predator.

Intertidal: that part of the shore which lies between high water and low water levels, being exposed at low tide and covered during high tide.

Invertebrate: an animal which lacks vertebrae which form a backbone.

Jaws: a pair of hard structures inside the mouth used for biting and/or holding.

Labile: unstable; such as chemicals liable to change; so a heat labile venom would be deactivated by heat.

Larva/larval: an early stage in the life cycle of many organisms, usually having a form and habit which is different from that of the adult stage.

Lappet: a tiny flap, as on the edge of the bell of some jellyfish.

Lesion: (1) wound, injury, or pathological change in the body (2) visible localised abnormality seen on the skin or body surface.

Ligature: A cord or thread used to tie around arteries in order to stop the circulation through them or to prevent escape of blood from their cut ends.

Maculotoxin: a toxin produced by the venom glands of octopus species belonging to the genus *Hapalochlaena*.

Mangroves: trees which are tolerant of saline water and so can grow in the intertidal zone, particularly along tropical and subtropical coasts.

Mantle: a flap or pocket in the skin of a mollusc under or in which the gills and/or other organs are located, and which may contain glands which secrete the materials forming the shell.

Mantle cavity: the space protected by the overlying mantle of a mollusc.

Medusa: the jellyfish stage in the life cycle of a coelenterate.

Melena: black stools indicating the presence of blood which has gone all the way through the intestines and been altered; the amount of blood lost is significant in assessing the severity of the cause of blood loss.

Metabolism: All the physical and chemical processes by which the living body is maintained.

Metabolite: A product of metabolism.

Micron: one thousandth of a millimetre (μm).

Morphology: the shape and arrangement of the whole body or organs of an organism.

Moult: discard of the exoskeleton of an arthropod allowing for further growth within a new and initially soft, flexible and expansible new exoskeleton.

Mouth parts: the paired structures surrounding the mouth of an arthropod which are used to hold, bite, suck and/or shred its food.

Mucus: a viscous, sticky or slimy fluid secreted internally or externally by glands.

Necrosis: the death of a limited portion of tissue; see also dermonecrosis.

Nematocyst: the stinging mechanisms within certain cells of jellyfish and other coelenterates.

Neurotoxin: a venom or poison which acts on the nervous system of a victim.

Noradrenaline: A precursor of adrenaline in the adrenal glands. Its main function is to mediate the transmission of impulses in the sympathetic nervous system. It also has a transmitter function in the brain.

Notochord: an unsegmented skeletal rod possessed, even temporarily, by all members of the phylum Chordata; it lies lengthwise between the central nervous system and the gut.

Nudibranchs: members of a large group of gastropod molluscs belonging to the subclass

Opisthobranchia; they are all hermaphrodites and do not have a shell to protect their gills or branchiae.

Ocellus: a little simple eye, incapable of forming an image but sensitive to variation in light intensity.

Oral arms: long appendages attached around the mouth of, for instance, a jellyfish.

Order: the usual major subdivision of a class or subclass, commonly comprising a plurality of families; a grouping of organisms more closely related to one another than to those in other orders within a class.

Osphradium: an organ within the mantle cavity of a mollusc which tests the chemical quality of the water drawn in by the gills.

Papain: a preparation derived from pawpaws (*Papaya*) which contains a peptidase enzyme; it attacks the peptide links between amino acids in a protein molecule, and so may also be called a proteinase.

Papilla: a small projection or outgrowth, usually on the skin (*pl.* papillae).

Paralysis: loss of muscle activity i.e. power.

Parapodium: one of a pair of muscular, sometimes bristly projections on either side of the segments of annelid worms (*pl.* parapodia).

Parasite: an animal which lives in or on another animal, obtaining nourishment at its host's expense but giving nothing in return.

Pedicellaria: a stalked pincer-like organ on the skin of some echinoderms which removes foreign particles and prevents the settling of fouling organisms (*pl.* pedicellariae).

Pedalium: the base of a tentacle or bunch of tentacles of a Cubomedusa.

Peptide: a chemical compound formed by the combination of two or more amino acid molecules.

Photosynthesis: the synthesis or manufacture of organic compounds from water and carbon dioxide using the energy of sunlight which is absorbed by the chlorophyll or green pigment found in most plants.

Phylum: a unit of classification below that of a kingdom; a grouping of organisms more closely related to one another than to those in other phyla within a kingdom.

Pinnule: Slender jointed appendages arising from each side of the arms in feather stars (crinoids). They carry minute tube feet which collectively form a food-gathering net.

Plankton: aquatic and generally small organisms which float or drift almost passively, usually near the water surface.

Poison: a toxin that enters the body via the digestive tract, through absorption through the intact skin.

Polyp: the sedentary form of a coelenterate which may reproduce sexually or asexually; the latter method may result in separate organisms or in colonies.

Polypeptide: a peptide formed by the union of three or more amino acid molecules.

Proboscis: an muscular extension of the lips which may project from the mouth or be inverted inside it when not in use.

Predator: a carnivorous animal (but not a parasite) which actively seeks and feeds upon other animals.

Protease: an enzyme which digests or breaks down protein molecules into more simple units.

Protein: a complex organic substance composed of numerous small amino acid molecules joined by peptide links to form the very large protein molecules.

Proteolytic venom: a venom containing an enzyme which causes the breaking down of proteins.

Radula: the chitinous tooth ribbon with its embedded rows of pointed teeth which is possessed by many types of molluscs; it covers a muscular 'tongue' which is turned outwards to rasp food; the complex radular teeth of cone shells are separate, not attached to a basal ribbon.

Respiration: the gas exchange of carbon dioxide from the body for oxygen from the air or water through the thin cell layer which forms the surface of gills or lungs.

Respiratory centre: an organ which controls the muscular movements involved in breathing in a vertebrate.

Reticulate: net-like.

Rigor: severe shivering.

Saliva: a secretion of mucus and digestive enzymes from the salivary glands into the mouth.

Saponin: a substance with soap-like frothing properties that is toxic or distasteful to predators of the animal or plant.

Saxitoxin: a toxin contained in the tissues of oysters and other molluscs; it is derived from toxins present in 'red tide' organisms.

Scavenger: an animal which feeds on the tissues of dead organisms, usually animals.

Sea hare: a member of the opisthobranch family Aplysiidae which has a pair of upstanding earlike sensory tentacles and a small fragile internal shell.

Sedentary: not mobile, may or may not be attached to the substrate.

Segment/ed: one of a repetitive set of sections along the length of the body which are fundamentally similar in structure.

Septicaemia: blood poisoning.

Sexual stage: the stage in the life cycle of an organism in which sexual reproduction occurs.

Shell: a hard outer covering of the body, usually refers to the calcareous outer skeleton of molluscs, but may also be used for the exoskeleton of arthropods.

Shoulder: that part of a gastropod shell which projects as a ridge between the shell aperture and the suture.

Silica (siliceous): a glassy inorganic substance.

Siphon: a muscular folded flap or tube extending from the edge of a mollusc's mantle; water may flow through a siphon into or away from the mantle cavity.

Skeletal muscles: striated muscles attached to the skeleton.

Snout: a projection from front of a mollusc's head which surrounds the mouth.

Sp.: an abbreviation for a single species of a particular genus.

Spp.: an abbreviation for more than one species of a particular genus.

Species: a taxonomic grouping of animals which are able to interbreed and to produce fertile young; species are grouped into genera.

Spicule: a sharp pointed rod or, in many cases, a complexly shaped structure strengthening the spongin skeleton of sponges or supporting the wall of sea cucumbers; it may consist of calcium carbonate or silica.

Spire: that part of a gastropod's shell which lies behind the shell aperture.

Spongin: a collagen-like protein forming the soft fibrous skeleton of sponges; it is usually strengthened by spicules of lime or silica.

Statocyst: an organ which detects changes in the position of an animal.

Striate: patterned with incised lines.

Substrate: the soft or hard substance or object to which an organism may attach itself or on which it may move e.g. the sandy bottom of a lake, a rocky reef, algal fronds, dead coral clumps, etc.

Suture: the zone of a gastropod shell which marks the line of fusion between one whorl and another.

Symbiosis: the intimate association of organisms of different groups, usually to the advantage of each; symbiotic unicellular algae in the tissues of corals, some jellyfish and other coelenterates provide carbohydrates to the coelenterate in return for nutrients the algae are unable to gather.

Systemic: relating to the entire organism rather than to any of its parts.

Tentacle: an elongate, usually muscular, sensory organ.

Test: the hard 'shell' of a echinoid (sea urchin).

Tetrodotoxin: a toxin contained in the tissues of bony fish of the family Tetraodontidae (toadfishes, puffer fishes).

Thermolabile: liable to chemical change if there is a change in temperature.

Tooth sac: a small pouch in which groups of the teeth of a cone are stored.

Toxin: a substance harmful to body tissues.

Umbrella: the 'body' of a scyphozoan jellyfish, which is also sometimes called a bell.

Urticaria: Intensely itchy raised weals on the skin which may be due to an allergy or a reaction to a sting.

Valve: an organ which allows the flow of a fluid in one direction but prevents any backflow.

Venom: a toxin that gains access to the body tissues by injection through the skin e.g. cone shell and jellyfish venoms.

Vesicle: a small blister or tube, or a small pouch opening from the side of a tube.

Visceral mass: that part of the body of a mollusc which includes the organs of digestion, excretion, reproduction etc but which does not include the head or the main muscular organs used for locomotion, such as the foot of a snail, the arms of an octopus, etc.

Weal: a raised white area of damaged skin with reddened margins.

Whelk: a common name for a member of some groups of carnivorous gastropods.

Whorl: a complete turn or revolution around the axis of a coiled gastropod shell.

Zooxanthellae: single-celled algae which may live symbiotically within the body of a host animal, such as the giant clams or many of the hard corals. They make use of carbon dioxide and nitrogenous wastes given off by the host and in return provide it with oxygen and synthesised food; at times the algae themselves may be digested by the host.

References

ALGAE

Anon, 1995, Outbreak of Gastrointestinal Illness associated with Consumption of Seaweed — Hawaii, 1994. *MMWR* **44**(39): 724–727.

Haddock, R., 1993, Guam seaweed poisoning: food histories. *Micronesica* **26**: 35–8.

Hallegraeff, G.M., 1993. A review of harmful algal blooms and their apparent global increase. *Phycologia* **32**: 79–99.

Jones, M.M. 1991. Toxic algae: 25–31. *In* Marine organisms transported in ballast water. A review of the Australian scientific position. *Bureau of Rural Resources, Bull.* No. 11. Canberra, Australian Government Publishing Services.

Nagai, H., Yasumoto, T. and Hokama, Y., 1996, Aplysiatoxin and debromoaplysiatoxin as the causative agents of a red alga *Gracilaria coronopigolia* poisonong in Hawaii. *Toxicon* **34**(7): 753–761

Noguchi, T., Matsui, T., & Miyazawa, K., 1994, Poisoning by the red alga 'Ogonori' (*Gracilaria verrucosa*) on the Nojima Coast, Yokohama, Kanagawa Prefecture, Japan. *Toxicon* **32**: 1533–8.

COELENTERATES

Azila, N.A., Othman, I. 1994. Cases of jellyfish envenomation in Malaysia. *Toxicon* **32**: 532.

Barnes, J.H. 1966. Studies on three venomous cubomedusae. In: *The Cnidaria and their evolution. Symp. zool. Soc. London* No. 16. (W.J. Rees, ed.): 307–332.

Bloom, D.A., Burnett, J. 1999. Effects of verapamil and CSL antivenom on *Chironex fleckeri* (Box jellyfish) induced mortality. *Toxicon* **37**(11): 1621–26.

Burnett, J.W. (ed.) 1999. International Consortium for Jellyfish stings – Jellyfish Sting Newsletter. (Supplement to ACTM Bulletin). See http://www.tropmed.org.au.htm

Burnett, J.W. and Calton, G.J. 1974. Sea nettle and man-o'-war venom: chemical comparison of their venoms and studies on the pathogenesis of the sting. *J. Invest. Dermatol.* **62**: 372–277.

Burnett, J.W. and Calton, G.J. 1977. The chemistry and toxicology of some venomous pelagic coelenterates. *Toxicon* **15**: 177–196.

Burnett, J.W., Rubinstein, H., Calton, G.J. 1983. First Aid for jellyfish envenomation. *Southern Medical Journal* **76**(7): 870–872.

Carrette, J.C., Cullen, P., Little, M., Periera, P.L. and Seymour, J.E. 2002. Temperature effects on box jellyfish venom: a possible treatment for envenomed patients? *Medical Journal of Australia* **177**(11/12): 654–655.

Corkeron, M.A. 2003. Magnesium infusion to treat Irukandji syndrome. *Medical Journal of Australia.* 178: 411.

Endean, R., Duchemin, C., McColm, D. and Fraser, E.H. 1969. A study of the biological activity of toxic material derived from nematocysts of the cubomedusan *Chironex fleckeri. Toxicon* **6**: 179–204.

Exton, D.R., Fenner, P.J. and Williamson, J.A. 1989. Cold packs: effective topical analgesia in the treatment of painful stings by *Physalia* and other jellyfish. *Medical Journal of Australia* **151** (4/18): 625–626.

Fautin, D.G. and Allen, G.R. 1997. *Anemone Fishes and their Host Sea Anemones.* Revised edition. Western Australian Museum, Perth. 160 pp.

Fenner, P.J. 1991. Cubozoan jellyfish envenomation syndromes and their medical treatment in northern Australia. *Hydrobiologia* **216/217**: 637–640.

Fenner, P.J. 1998. Management of Marine Envenomation: Pt 1. Jellyfish. *Modern Medicine of Australia* **1998**(1): 22–28.

Fenner, P.J. 2005a. Dangerous Australian box jellyfish. *South Pacific Underwater Medicine Society (SPUMS) Journal.* **35**(2): 79–83.

Fenner, P.J. 2005b. Venomous jellyfish of the world. *South Pacific Underwater Medicine Society* (SPUMS) *Journal.* **35**(3): 131–138.

Fenner, P.J. and Carney, T. 1999. The Irukandji Syndrome. A devastating syndrome caused by a north Australian jellyfish. *Australian Family Physician* **28**(11): 1131–1137.

Fenner, P.J., Fitzpatrick, P.F. 1986. Experiments with the nematocysts of *Cyanea capillata. Medical Journal of Australia.* **145**: 174.

Fenner, P.J. and Hadok, J.C. 2002. Fatal envenomation by jellyfish causing Irukandji syndrome. *Medical Journal of Australia.* **177**: 362–363.

Fenner, P.J., Heazlewood, R.J. 1999. Papilloedema and coma in a child: undescribed symptoms of the Irukandji Syndrome. *Medical Journal of Australia.* **167**: 650–651.

Fenner, P.J., and Lewin, M. 2003. Sublingual glycerol trinitrate as prehospital treatment for hypertension in Irukandji syndrome. *Medical Journal of Australia.* **179**: 655.

Fenner, P.J., Williamson, J.A., Burnett, J.W., Colquhoun, D.M., Godfrey, S., Gunawardane, K. and Murtha, W. 1988. The 'Irukandji Syndrome' and acute pulmonary oedema. *Medical Journal of Australia.* **149**: 150–155.

Fenner, P.J., Williamson, J.A., Burnett, J.W., Rifkin, J. 1993. First aid treatment of jellyfish stings in Australia: response to a newly differentiating species. *Medical Journal of Australia.* **158**: 498–501.

Freeman, S. E. and Turner, R. J. 1972. Cardiovascular effects of cnidarian toxins: a comparison of toxins extracted from *Chiropsalmus quadrigatus* and *Chironex fleckeri*. Toxicon **10**: 31–37

Gershwin, L. 2005a. Two new species of jellyfish (Cnidaria: Cubozoa: Carnbdeida from tropical Western Australia, presumed to cause Irukandji Sundrome. *Zootaxa* **1084**: 1–30

Gershwin, L. 2005b. *Carybdea alata* auct. and *Manokia stiasnyi*, reclassification to a new famiy with description of a new genus and two new species. *Memoirs of the Queensland Museum* **51**(2): 501–523.

Gershwin, L. 2006b. Nematocysts of the Cubozoa. *Zootaxa* **1232**: 1–57.

Gershwin, L.A. 2006a. Comments on *Chiropsalmus* (Cnidaria; Cubozoa; Chirodropida): a preliminary revision of the Chiropsalmidae, with descriptions of two new genera and two new species. *Zootaxa* **1231**: 1–42.

Gershwin, L.A. and Zeidler, W. 2008. Some new and previously unrecorded Scyphomedusae (Cnidaria; Scyphozoa) from southern Australian coastal waters. *Zootaxa* **1744**: 1–18.

Gleibs, S., Mebs, D. 1999. Distribution and sequestration of palytoxin in coral reef animals. *Toxicon*: **37**(11): 1521–1527.

Hadok, J.C. 1997. Irukandji Syndrome: a risk for divers in tropical waters. *Medical Journal of Australia.* **167**: 649–650.

Hamner, W.M., Jones, M.S. and Hamner, P.P. 1995. Swimming, feeding, circulation and vision in the Australian box jellyfish, *Chironex fleckeri* (Cnidaria: Cubozoa) *Mar. Freshw. Res.* **46**: 985–990.

Hartwick, R. 1987. The box jellyfish pp. 98–105: *In*: J. Covacevich, P. Davie and J. Pearn (eds): *Toxic plants and animals: a guide for Australia.* Queensland Museum, Brisbane. 504pp.

Hartwick, R.F. 1991. Distributional ecology and behaviour of the early life stages of the box-jellyfish *Chironex fleckeri. Hydrobiologia* **216/217**: 181–188.

Hartwick, R.J., Callanan, V., Williamson, J.A.H. 1980a. Disarming the box jellyfish – nematocyst inhibition in *Chironex fleckeri. Medical Journal of Australia* 1980 **1**: 15–20.

Hartwick, R.J., Callanan, V., Williamson, J.A.H. 1980b. Disarming the box jellyfish (letter in reply). *Medical Journal of Australia* **1**: 335–338.

Huynh, T.T., Seymour, J., Periera, P., Mulcahy, R., Cullen, P., Carrette, T. and Little, M. 2003. Severity of Irukandji stings and nematocyst identification from skin scrapings. *Medical Journal of Australia* **178**: 38–41.

Little, M., Mulcahy, R.E. 1998. A year's experience of Irukandji envenomation in far north Queensland. *Medical Journal of Australia* 1998. **169**: 638–641.

Loten, C., Strokes, B., Worsley, D., Seymour, J. E., Jiang, S., and Isbister, G. K. 2006. A randomised controlled trial of hot water (45°C) immersion versus ice packs for pain relief in bluebottle stings. *Medical Journal of Australia* **184**(7): 329–333.

Makrokanis, C.J., Hall, N.L. and Mein, J.K. 2004. Irukandji syndrome in northern Western Australia: an emerging health problem. *Medical Journal of Australia* 2004. **181**(11/12): 699–702.

Maretić, Z., Matic-Plantanida, D., Ladarc, J., 1987. The bloom of the jellyfish *Pelagia noctiluca* in the Mediterranean and Adriatic and its impact on human health. pp. 260–267: *In*: Proceedings of the 2nd workshop on jellyfish in the Mediterranean Sea, Trieste. 2–5 Sept. 1987. Mediterranean Action Plan of the United Nations Environmental Programme, Athens, 1991.

Martin, C. and Audley, I. 1990. Cardiac failure following Irukandji envenomation. *Medical Journal of Australia* 1990. **153**: 164–166.

Mebs, D. and Gebauer, E. 1980. Isolation of proteinase inhibitory, toxic and hemolytic polypeptides from a sea anemone *Stoichactis* sp. *Toxicon* **18**: 97–106.

Middlebrook, R.E., Wittle, L.W., Scura, E.D. and Lane, C.E. 1971. Isolation and purification of a toxin from *Millepora dichotoma. Toxicon* 9: 333–336.

Moore, R.E. and Scheuer, P.J. 1971. Palytoxin: a new marine toxin from a coelenterate. *Science* **172**: 495–498.

Neeman, I., Calton, G.J. and Burnett, J.W. 1980. Cytotoxicity and dermonecrosis of sea nettle *(Chrysaora quinquecirrha)* venom. *Toxicon* **18**: 55–63.

Othman, I., Aini, Y., Yusuff, A.W., Burnett, J.W. and Azila, N. 1991. Biochemical studies of *Chironex fleckeri* and the treatment of *Chironex* envenomation with traditional herbal preparations. *Toxicon* **28**: 821–835.

O'Reilly, G.M., Isbister, G.K., Treston, C.T., Lawrie, P.M. and Cujrrie, B.J. 2001. Prospective study of jellyfish stings from tropical Australia, including the major box jellyfish *Chironex fleckeri. Medical Journal of Australia* **175**: 652–655.

Patocka, J. and Strunecka, A. 1999. Sea anemone toxins, *The ASA Newsletter*, Article 99.1b. 3 pp. (Applied Science and analysis, Inc.)

Pongrayoon, U., Bohlin, L. and Wasuwat, S. 1991. Neutralization of toxic effects of different crude jellyfish venoms by an extract of *Ipomoea pes-caprae* (L.) R. Br. *Journal of Ethnopharmacology* **35**: 65–69.

Radwan, F.F.Y., Burnett, J.W., Bloom, D.A., Coliano, T., Eldefrawi, M.E., Erdely, H., Aurelian, L., Torres, M. and Heimer-de la Cotera, E.P. 2001. A comparison of the toxinological characeristics of two *Cassiopea* and *Aurelia* species. *Toxicon* **39**(2–3): 245–257.

Rifkin, J.F., Fenner, P.J. and Williamson, J.A. 1993. First aid treatment of the sting from the hydroid *Lytocarpus philippinus*: the structure of, and in vitro discharge experiments with its nematocysts. *Journal of Wilderness Medicine*. **4**: 252–260.

Rifkin, J.F. 1996. Jellyfish mechanisms. Ch. 6 *in* Williamson, J.A., Fenner, P.J., Burnett, J.W. and Rifkin, J.F. (eds) 1996. *Venomous and poisonous marine animals: a medical and biological handbook.* University of NSW Press, Sydney: 121–170.

Seymour, J. and Sutherland, P.A. 2001. Box Jellies. *Nature Australia*, Autumn 2001: 32–41.

Shryock, J.C. and Bianchi, C.B. 1983. Sea nettle (*Chrysaora quinquecirrha*) nematocyst venom: mechanism of action on muscle. Toxicon **21**(1): 81–95.

Southcott, R.V. 1956. Studies on Australian Cubomedusae, including a new genus and species apparently harmful to man. *Aust. J. Mar. Freshw. Res.* 7: 254–280.

Southcott, R.V. 1967. Revision of some Carybdeidae (Scyphozoa: Cubomedusae), including a description of the jellyfish responsible for the 'Irukandji syndrome'. *Aust. J. Zool.*, **15**: 651–671.

Tardent, P. 1997. How do cnidaria make use of their venomous stinging cells? *Toxicon* **35**: 818.

Taylor, G. 2000. Are some jellyfish toxins heat labile? *South Pacific Underwater Medicine Society (SPUMS) Journal* **30**(2): 74–75.

Taylor, J.G. 2007. Treatment of jellyfish stings. Letter to editor, *The Medical Journal of Australia* **186**(1): 43.

Williamson, J.A., Callanan, V.I. and Hartwick, R.F. 1980. Serious envenomation by the northern Australian box-jellyfish *(Chironex fleckeri). Medical Journal of Australia* 1980 **1**: 13–15.

Wittle, L.W., Middlebrook, R.E. and Lane, C.E. 1971. Isolation and partial purification of a toxin from *Millepora alcicornis. Toxicon* **9**: 327–331.

Wittle, L.W., Scura, E.D. and Middlebrook, R.E. 1974. Stinging coral (*Millepora tenella*) toxin: a comparison of crude extracts with isolated nematocyst extracts. *Toxicon* **12**: 481–486.

Wittle, L.W. and Wheeler, C.A. 1974. Toxic and immunological properties of stinging coral toxin. *Toxicon* 12: 487.

CRUSTACEA

Llewellyn, L.E. and Endean, R. 1991. Paralytic shellfish toxins in the xanthid crab *Atergatis floridus* collected from Australian coral reefs. *J. wilderness Med.* **2**: 118–126.

Pritchard, M.1991. Lice attacks spark fear for swimmers. *The West Australian* 13 May 1991: 9.

Raj, U., Haq, H., Oshima, Y. and Yasumoto, T. 1983. The occurrence of paralytic shellfish toxins in two species of xanthid crab from Suva Barrier Reef, Fiji Islands. *Toxicon* **21**(4); 547–551.

ECHINODERMS

Alender, C.B. and Russell, F.E. 1966. Pharmacology. In: *Physiology of Echinodermata*, (R.A. Boolootian, ed.): 529–543. *Interscience*: New York.

Barnes, J. and Endean, R. 1964. A dangerous starfish –– *Acanthaster planci* (Linnaeus). *Medical Journal of Australia* 1964 **1**: 592–593.

Earle, K.V. 1941. Echinoderm injuries in Nauru. *Medical Journal of Australia* 1941 **2**: 265–266.

Endean, R. 1961. The venomous sea-urchin *Toxopneustes pileolus. Medical Journal of Australia*, 1961, 1: 320.

Feigen, G.A., Sanz, E. and Alender, C.B. 1966. Studies on the mode of action of sea urchin toxin. I. Conditions affecting release of histamine and other agents from isolated toxins. *Toxicon* **4**: 161.

Friess, S.L. 1963. Some pharmacological activities of the sea cucumber neurotoxin. *American Institute of Biological Sciences Bulletin* **13**(2): 41.

Kimura, A. and Nakagawa, H. 1980. Action of an extract from the sea urchin, *Toxopneustes pileolus*, on isolated smooth muscle. Toxicon **18**: 689–693.

Mebs, D. 1984. A toxin from the sea urchin *Tripneustes gratilla. Toxicon* **22** (2): 306–307.

Mortensen, T. 1935–1943. *A monograph of the Echinoidea*. Vols II, III. C.A. Reitzel: Copenhagen.

Nakagawa, H., Tu, A.T. and Kimura, A. 1991. Purification and characterization of contractin A from the pedicellarial venom of sea urchin *Toxopneustes pileolus – Archives of Biochemistry and Biophysics*. **284**(2): 279.

Pope, E.C. 1964. A stinging by a crown-of-thorns starfish. *Australian Museum Magazine*. **14**: 350.

Pope, E.C. 1968. Venomous starfish in Sydney Harbour. *Australian Natural History*. **16**(1): 26.

Saito, T. and Kishimoto, H. 2003. Tetrodotoxin attracts toxic starfish, *Astropecten polyacanthus. Bull. Inst. Oceanic Research and Development*, Tokai University **24**: 45–49.

Wilson, B.R. and Stoddart, J. 1987. A thorny problem Crown-of-Thorns starfish in WA. *Landscope* spring 1987: 35–39.

GENERAL

Catala, R. 1964. *Carnival under the sea*. R. Sicard: Paris. 141 pp.

Cleland, J.B. and Southcott, R.V. 1965. *Injuries to man from marine invertebrates in the Australian region.* National Health and Medical Research Council, Special Report Series No. 12. Canberra. 282 pp.

Covacevich, J., Davie, P. and Pearn, J. (eds) 1987. *Toxic Plants and Animals. A guide for Australia.* Queensland Museum, Brisbane. 504 pp.

Cribb, J. 1985. Australian Medicinal Plants. *Medical Journal of Australia.* **143**: 574–577.

Edmonds, C. 1981. *Marine animal injuries to man.* Wedneil Publications, Melbourne/Newport.

Edmonds, C. 1984. *Marine animal injuries to man.* Wedneil Publications, Melbourne/Newport.

Edmonds, C. 1989. *Dangerous marine creatures.* Reed Books Pty Ltd Frenchs Forest, NSW. 192 pp.

Fenner, P. 1987. Marine envenomations. *Australian Family Physician.* **16**(2): 93–102.

Halstead, B.W. 1965. *Poisonous and venomous marine animals of the world* 1: U.S. Govt Printing Office: Washington D.C.

Halstead, B.W. 1995. *Dangerous Marine Animals: that bite, sting, shock, or are non-edible.* 3rd ed. Cornell Maritime Press, Centreville: i-x, 1–275.

Hashimoto, Y. 1977. *Marine toxins and other bioactive marine metabolites.* Japan Scientific Societies Press: Tokyo. [English edition, 1979]. 369 pp.

Pearn, J. and Covacevich, J. 1988. *Venoms and Victims.* The Queensland Museum and Amphion Press. 135 pp.

Russell, F.E. 1984. Marine toxins and venomous and poisonous marine plants and animals (Invertebrates). *Adv. Mar. Biol.* **21**: 60–194.

Russell, F.E. Nagabhushanam, R. 1996. *The venomous and poisonous marine invertebrates of the Indian Ocean.* Science Publishers, Inc, Enfield, New Hampshire, i-vii, 1–271.

Sutherland, S.K. 1983. *Australian animal toxins.* Oxford University Press: Oxford, Auckland, New York.

Sutherland, S.K. and Tiballs, J. 2001. *Australian animal toxins.* Oxford University Press: Melbourne. 2nd Edition.

Williamson, J.A., Fenner, P.J., Burnett, J.W. and Rifkin, J.F. (eds) 1996. *Venomous and poisonous marine animals: a medical and biological handbook.* University of NSW Press, Sydney: 1–504.

MOLLUSCS

Beesley, P.L., Ross, G.J.B. & Wells, A. (eds) 1998. *Mollusca: The southern Synthesis. Fauna of Australia.* Vol. 5. CSIRO Publishing: Melbourne, Part A, xvi, 563 pp.

Kohn, A.J., (1998) Superfamily Conoidea: 846-854 *In:* Beesley, P.L., Ross, G.J.B. & Wells, A. (eds) *Mollusca: The Southern Synthesis . Fauna of Australia. Vol. 5.* CSIRO Publishing: Melbourne, Part B viii 565–1234 pp. : 852.

Kohn, A.J., 2003, The feeding process of *Conus victoriae* : 101–107. *In*: Wells, F. E., Walker. D.I, and Jones D.S. (eds) *The marine flora and fauna of Dampier, Western Australia.* Western Australian Museum, Perth.

Melo, V.M.M., Duarte, A.B.G., Carvalho, A.F.F.U., Siebra, E.A. and Vasconcelos, I.M. 2000. Purification of a novel antibacterial and haemagglutinating protein from the purple gland of the sea hare, *Aplysia dactylomela* Rang, 1828. *Toxicon* **38**(10): 1415–1427.

Messenger, J.G. 2001. Cephalopod chromatophores; neurobiology and natural History. *Biol. Rev.* **76**: 473–528.

Norman, M. and Reid, A. 2000. *A guide to squid, cuttlefish and octopuses of Australasia.* The Gould League of Australia, Moorabin, Vic. and CSIRO Publishing, Collingwood, Vic. 96 pp.

Wang, Y., Yu, J., Yan, T., Fu, M. & Zhou, M., 2003, Advance of research on application of Paralytic shellfish Poisoning toxins. *Studia Marina Sinica* No. 45: 124–131

Wilson, B.R. (1994) *Australian Marine Shells. Prosobranch Gastropods Pt 2 (Neogastropods).* Odyssey Publishing : Kallaroo, Western Australia. 370 pp.

SPONGES

Bergquist, P. and Skinner, I.G. 1982. Porifera pp. 38–72. *In*: *Marine invertebrates of southern Australia.* (S.A. Shepherd and I.M. Thomas, eds). South Australian Govt Printer: Adelaide.

Hooper, J.N.A., Capon, R.J. and Hodder, R.A. 1991. A new species of toxic marine sponge (Porifera: Demospongiae: Poecilosclerida) from northwest Australia. *The Beagle, Records of the Northern Territory Museum of Arts and Sciences* **8**(1): 27–36.

Southcott, R.V. and Coulter, J.R. 1971. The effects of the southern Australian marine stinging sponges, *Neofibularia mordens* and *Lissodendoryx* sp. *Medical Journal of Australia* 1971 **2**: 895–901.

SWIMMERS' ITCH

Appleton, C.C. and Lethbridge, R.C. 1979. Schistosome dermatitis in the Swan Estuary, Western Australia. *Medical Journal of Australia* **1**: 141–144.

Bearup, A.J. 1956. Life cycle of *Austrobilharzia terrigalensis.* Johnston, 1917. *Parasitology* **46**: 470–479.

Pope, E.C. 1972. Terrigal itch, a coastal waters dermatitis. *Australian Natural History* **17**(7): 217–221.

Index

A

B

C

D

E

About the Authors

Loisette Marsh graduated from the University of Western Australia in 1956 with a MA in Zoology after research on the fauna of intertidal rock platforms.

As a scuba diver from 1958 to 1994 she dived, snorkelled and reef walked on coral reefs from the Tuamotus to Mauritius, including a four-year stint in Fiji and participation in one of the Indonesian Rumphius expeditions to the Maluccas in eastern Indonesia.

During 23 years in the Department of Aquatic Zoology at the Western Australian Museum she took part in dive surveys of Western Australian coral reefs, from the Abrolhos to the Kimberley, the shelf-edge atolls (Rowley Shoals, Scott and Ashmore Reefs) as well as Christmas and Cocos (Keeling) islands. Taxonomy and distribution of corals and echinoderms were her main focus in these surveys but since retirement in 1993 she has concentrated on echinoderms, sea star taxonomy in particular.

During her long association with coral reefs she has seen and safely handled most of the venomous animals featured in this book and been stung by some, but never seriously.

Loisette co-authored the first edition of Sea Stingers (1986) and Hermatypic Corals of Western Australia (1988) as well as taxonomic papers on sea stars and numerous coral reef fauna survey reports.

Shirley Slack-Smith graduated from the University of Sydney in 1956, and after teaching secondary school science in Brisbane and Victoria, and working in various research capacities for the Museum of Victoria, the University of Western Australia and the Western Australian Museum, she was appointed Curator of Molluscs at the Western Australian Museum in 1971, a position she still holds.

During her years with the museum she has carried out extensive field research in various parts of Australia and overseas, including in Papua New Guinea, Malaysia, Thailand, the Philippines, Taiwan, India, Sri Lanka, Madagascar and Mauritius.

Published 2010 by the
Western Australian Museum
49 Kew Street, Welshpool, Western Australia 6106
www.museum.wa.gov.au

First edition published 1986.

Designer Cathie Glassby
Printed by Everbest Printing, China.

National Library of Australia
Cataloguing-in-Publication entry
Author: Marsh, Loisette M.
Title: Field guide to sea stingers and other venomous and poisonous marine invertebrates in Western Australia / Loisette M. Marsh, Shirley M. Slack-Smith.
Edition: 2nd ed.

ISBN: 9781920843953 (pbk.)

Notes: Includes index. Bibliography.
Subjects: Poisonous marine invertebrates — Western Australia.
Marine invertebrates — Western Australia.
Dangerous marine animals — Western Australia.
Other Authors/Contributors: Slack-Smith, Shirley M.
Western Australian Museum.

Dewey Number: 593.09941

Front Cover: *Pachycerianthus* sp. (Clay Bryce).
Back Cover (from the top): *Diadema* spp. (Clay Bryce); *Hapalochlaena lunulata* (Clay Bryce); *Versuriga anadyomene* (Roger Steene).